AUSTRALIS OSCAR 5

OWEN MACE AND RICHARD TONKIN

ATF
SPACE
Expanding Horizons
ADELAIDE
2019

AUSTRALIS OSCAR 5

The improbable story of Australia's first private satellite.

ATF
SPACE

Expanding Horizons

ADELAIDE

2019

National Library of Australia Cataloguing-in-Publication entry

Creator: Mace, Owen & Tonkin, Richard (authors).
Title: Australis OSCAR 5 : the story of how Melbourne University
               students built Australia's first satellite / Owen Bullivant Mace.

ISBN: 978-1-925612-41-7 (soft)
               978-1-925612-42-4 (hard)
               978-1-925612-43-1 (epub)
               978-1-925612-44-8 (PDF)

Subjects: Melbourne University students.
Artificial satellites--Australia.
Artificial satellites--Launching.

An imprint of the ATF Ltd
PO Box 504 Hindmarsh
SA 5007
ABN 90 116 359 963
www.atfpress.com

Graphic Design & Layout: Lydia Paton
Font (Body text): Garamond (11.5pt)
Font (Heading text): Garamond (28pt), Futura LT (12pt)

To the young students who built and flew Australis, and to the memory of the late Geoff Thompson and Les Jenkins—they reached for the stars.

OWEN MACE AND RICHARD TONKIN

Acknowledgements

Australis was a team effort across the continents, first and foremost. Some are mentioned in this book and some, unfortunately, not. Thanks to each and very one of the Australis, WIA, OSCAR, AMSAT and NASA teams and to the many amateur radio operators throughout the world who listened to its beepings. Without each and every one, Australis OSCAR 5 would not have flown and not been the undoubted success that it was. Each person contributed something to the project and was part of its success. It seems unfair to name those I remember and leave out those whose names have slipped from my memory. They all contributed and thanks to them all.

Credit is due to Dr. Alice Gorman ('Dr. Space Junk') for her suggestion to write a book about Australis and to Anne Johnson for her documentary film that was my inspiration.

I must acknowledge my publisher, Hilary Regan and ATF Press, for taking on the task of steering me, who has never written a book before, through the publishing maze; for the patience of the copy-editor, Amy Lovat, for finding literally thousands of errors and changes necessary to make my initial manuscript readable and to Lydia Paton for her beautiful layout of the book. Thanks, too, to Pauline and Richard for their meticulous checking.

My English teachers deserve credit, too, for their patience with a student whose skill with the written word can only be judged by the 'rivers of red ink' that covered my essays at school.

And thanks to my beloved wife, Delia, for her decades of patience with her nerdy husband and for her wise counsel.

Owen Mace
ADELAIDE, AUGUST 2017

Plate 1a *This is a publicity photograph of the team taken in a Carlton alley very near the University of Melbourne. We are looking a good deal younger in the photo than today! In the top row, left to right, are: John, Paul, Richard (behind the antenna) and Geoff. Kneeling in the drain are Owen, Peter and Steve.*

Contents

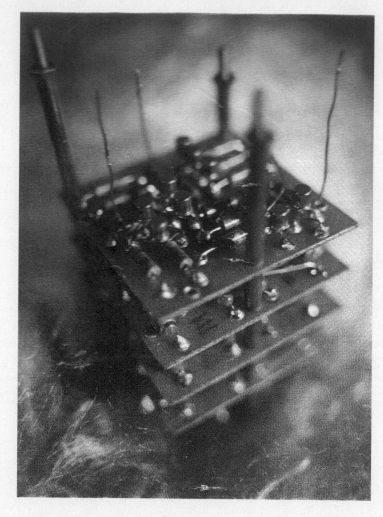

What is this? Read on to find out...

Plate 1b *What is this? Read on to find out.*

ix   AUSTRALIS OSCAR 5

Introduction

This is the story of the first satellite built in Australia, one that was made by a group of students at the University of Melbourne in the 1960s and launched by NASA in 1970. We have been told that our achievement was remarkable and yet we did not think so at the time. We were simply following our interest in anything space. Nevertheless, we all recall those times with pleasure and pride. What we did achieve was unusual, although that was not our intention. We just wanted to do something that fascinated us and the fact that no one in Australia had done anything like it before was of no concern or interest to us.

As this is being written, the 50th anniversary of the first of the moon landings has recently taken place. It will soon be the 50th anniversary of the launch of Australis, almost exactly six months after Neil Armstrong stepped onto the moon. Two vastly different projects; one immense and requiring the full attention of the mightiest country on earth, the other tiny. And yet there are parallels: both required the participants to acquire new knowledge, be very dedicated and both achieved many firsts. A few of those NASA people who had achieved so much turned to the issues of launching a satellite built by foreign students. We have much to thank them for.

This all happened over 50 years ago, and some of our memories have dimmed a little, or rather have dimmed a lot, and there are even instances where we disagree. Where there are differences, they are highlighted. Does it matter now whether our memories of details are the same? Probably not. Our activities and contributions to the project differed and our memories likewise differed.

The first edition of this book[1] recorded our version of a singular event in the lives of a disparate group of mainly undergraduate students at the University of Melbourne from 1965 to 1970 and beyond. We were bound together by a fascination with anything and everything relating to rockets, satellites, lunar and interplanetary probes.

As if to recognise our launch, another group of University of Melbourne students have recently had their satellite, ACRUX-1, launched from New Zealand. Their story finds a place in this second edition.

What is remarkable are the changes that have taken place since then and emphasised by the ACRUX chapter. The developments in technology, especially electronics and software, are obvious and incredible. Huge social changes, too, and vastly increased regulation in all parts of our lives, including space.

* * *

---

[1] Australis OSCAR 5, 2017, ATF Press, ISBN 9781925309812. It was written in time for the International Astronautical Congress held in Adelaide, South Australia in September 2017

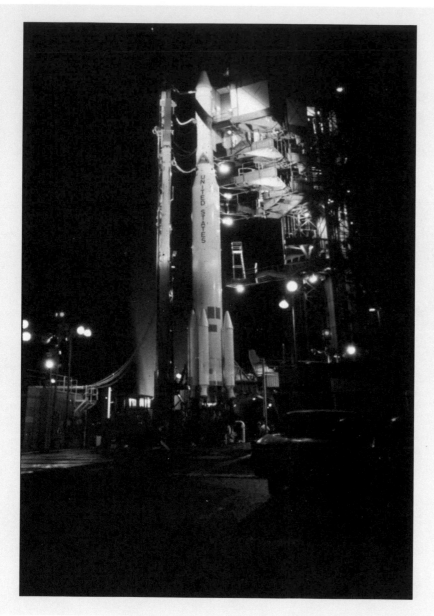

*Delta 76
with Australis
(soon to become
Australis OSCAR 5)
on the launch pad
ready to fly!*

Plate 1c *Delta 76 with Australis (soon to become Australis OSCAR 5) on the launch pad ready to fly*

It should be recorded that Richard and Owen attended the 50th Anniversary Symposium of AMSAT, the organisation that arranged the launch of Australis OSCAR 5, in Washington, D.C., on October 18 and 19, 2019, where they presented a paper and spoke at the Symposium banquet. They were treated as honoured guests and met some of the American space pioneers mentioned in this book.

No similar event or recognition of the 50th anniversary of our satellite, on 23 January 2020, has yet materialised in Australia, other than this second edition, planned to be launched on that 50th anniversary. We hope that we will not be saying that there was more interest in this significant event in the US than in Australia.

* * *

The form of this book is based on the web site, www.australis-oscar5.weebly.com. Images have been credited, as appropriate. Photographs with no credit were taken by Australis team members.

Plate 1.1 *Sputnik 1. Image Credit: NASA/Asif A. Siddiqi, from https://www.nasa.gov/multimedia/imagegallery/image_feature_924.html* Accessed 31 July 2017.

CHAPTER 1
The Spark

'Pens down and onto the lawns' said "Frosty" Coleman.

Now Frosty was not someone to be trifled with, especially as he was the housemaster in a boys' boarding school, Geelong Grammar at Corio, and Owen was one of his eleven-year-old charges. Not only was Frosty a somewhat severe man, to be not so much feared as respected, in a fearful sort of way. He had been a bomber pilot in World War II, so he was certainly respected by his charges for his wartime exploits but also for the way he wielded his cane on boys who strayed from the designated path.

Those boys were born around the end of World War II and their parents lived through it. They were enthralled by wartime exploits, having read books telling the stories of the Dam Busters, Douglas Bader in Reach for the Sky and The Great Escape.

On this evening around 8pm, Frosty pointed to the lawn and playgrounds outside our study room for thirty or so boys. They trooped out, happy to be relieved of some study time and also a little fearful of what the reason might be. Perhaps it was something to do with Frosty's interest in the stars and astronomy? It was.

Once gathered on the playground grass on the cold, dark but clear night, Frosty commanded the boys in his bomber-pilot way to look up to the north west. Nothing but a thousand – no, a million – stars. And then something magical happened.

A spark of light was moving toward the group and in a couple of minutes it was overhead. Then it disappeared. That, Frosty explained, was an artificial earth satellite, the first ever. But not one of ours. It was the enemy's, the Soviet's. Would it attack? We wondered whether we were we doomed. Possibly not, because it had a funny and unfamiliar name: Sputnik.[1]

Owen was one of those boys and that spark of light lit a lifelong fascination in him with anything and everything about space. There were others, too, who saw the same thing and were equally as excited by that little spark.

This is the story of where that spark and their imaginations led them – into space.

Richard's introduction to space was remarkably similar to Owen's. Here is his story.

* * *

---

[1]   Crikey, 'Sputnik 1: the night Australians watched the skies' at <https://World Warw.crikey.com.au/2007/10/03/sputnik-1-the-night-australians-watched-the-skies/>. Accessed June 2017

RICHARD'S INTRODUCTION TO SPACE

My introduction to space research was remarkably similar, in fact almost identical, to Owen's. Like Owen, I was at boarding school when, on the night of 5 October 1957, our housemaster herded we Year 9 students up the hill to the Shrine War Memorial in Melbourne and told us to look up at the sky where we would see Sputnik, the first artificial satellite, which the Soviet Union had launched the day before. I suppose I had heard that they and the Americans had announced plans to orbit satellites during the International Geophysical Year, but the mere announcement had not piqued my interest.

But here was the real thing – a 'star' that moved quite quickly, it seemed, across the sky – an event that would have had its parallels in my grandparents being amazed at the invention of electricity, the Wright Brothers' first flight, television and the motor car.

I was 13 at the time – half child, half man – with nowhere else to direct my boundless adolescent energy. While my boarding school friends were tinkering with model aeroplane engines, playing sport or thinking of girls (which was all we were allowed to do), I soaked up as much information as I possibly could about the Russian and American space programmes. I had a small transistor radio that I would listen to wherever I could, for the latest space news – including in bed at night, where I pressed it to my ear with the blankets pulled over my head.

I took this to the extreme. The United States Air Force began a programme in 1959 called Discoverer (also referred to as Corona), which, as Owen says in Chapter Three, was, we thought, going to launch Australis in 1967. Discoverer was a spy satellite, with a capsule designed to be ejected from the parent craft and recovered as it parachuted to earth, by being snagged with a hook trailing from a small-moving piston engined aircraft. Through 1959 and 1960, they tried and they tried to make it work, but 11 times it failed – the launch rocket did not get Discoverer into orbit, or the capsule was ejected the wrong way. The failures went on and on. Now, in those days, space was still big news and every launch was reported by the media. The success or failure of the capsule recovery was announced on the 12.30pm ABC News. On the day of each attempt, I would excuse myself from my Maths class (which was no great loss, as I was, unlike Owen and the others, quite hopeless at the subject), saying that I had to go to the toilet, where I produced my little radio and listened anxiously to the news. This happened frequently enough for the Maths teacher to take me aside and ask whether I was getting enough All Bran for breakfast.

At long last, on a wet August day in 1960, I was perched on the loo, awaiting the fate of Discoverer 13. The ABC News' grand piece of band music played (as it still does today), heralding the 12.30pm bulletin. The first item was all that I heard: 'The United States today recovered a capsule from an orbiting satellite. An Air Force aircraft failed to snag it as it parachuted toward the Pacific Ocean south of Hawaii, but a navy diver, dropped from a helicopter, located the capsule bobbing in the sea and it was winched aboard. President Eisenhower has hailed the feat as a major advance in space exploration', or words to that effect.

I was ecstatic – I whooped 'Yes!' very loudly. A boy in the next cubicle must have clearly heard me and my little secret was out.

Now you might say that it was completely irrational for a then-16-year-old to get excited about another country's space success, but that was the way I was. The space bug infected me that October night when I first saw Sputnik and it is with me still, 60 years later.

So far as I know, I was the only one at school, day boys or boarders, who was at all interested in space. I was regarded as a nerd, because, once the initial excitement over Sputnik had faded away, for my school friends and the public generally, it was only to be rekindled in the early 1960s by the first human spaceflights, whereas I continued to be enthralled with each launch.

On my exeats, a couple of weekends a term when we boarders were allowed to go home, and on school holidays, I would build two stage rockets out of skyrockets that could then still be bought freely from fireworks shops in China Town, which occupied a couple of blocks in Little Bourke Street in Melbourne. The parallels with Owen's similar efforts, are remarkable. I launched these space vehicles from the backyard at home, usually with less than satisfactory results: many fell into neighbouring gardens, spluttering and smoking. Luckily, our neighbours were sympathetic to 'the Tonkin kid' and his strange hobby. They called it a hobby, to me it was an obsession.

* * *

We now return to the lead up to Sputnik 1 and events following it.

The International Geophysical Year (IGY, 1 July 1957 to 31 December 1958) was eighteen months of co-operative international scientific study of the Earth that had huge significance. Sixty seven countries, including Cold War enemies, The Soviet Union and America, contributed to the advancement of knowledge about the Earth that led to the discovery of mid-ocean ridges and hence the theory of plate tectonics, as well as establishing Arctic and Antarctic research stations. The Antarctic Treaty committed countries to the peaceful and scientific uses of the Antarctic and was a direct result of the IGY. Sputnik was launched during the IGY and its role was to explore the electron content of the ionosphere. There is a great deal of detailed information about Sputnik 1 in Wikipedia.[2]

In the meantime Owen scoured every newspaper for the slightest detail, and there was almost none. He recalls one of the annual trips into 'town' – the central business district of Melbourne – with his grandmother. The headlines of the afternoon newspaper shrieked that there was now a second Sputnik and, evidently, a great deal bigger and heavier than the first one.[3] The Soviets had launched a second Sputnik[4] on 3 November 1957 and it carried a dog, Laika, which died soon after in space. How sad that

---

2   Wikipedia, 'Sputnik 1' at <https://en.wikipedia.org/wiki/Sputnik_1>. Accessed June 2017.

3   Wikipedia, 'Sputnik 2' at <https://en.wikipedia.org/wiki/Sputnik_2>. Accessed June 2017.

4   Crikey, 'Sputnik 2: the space age Australia never had' at <https://World Warw.crikey.com.au/2007/10/03/

the dog had to die for the Soviets to prove their superiority. Much later, we learned that she had died a rather cruel death soon after reaching orbit from overheating and carbon dioxide suffocation.

To much fanfare and publicity, America announced that their first satellite was to be launched soon after Sputnik 2. The launch was televised and, much to our dismay, the first stage rocket sank after a few seconds and disappeared in an enormous ball of flame. Dismay because it seemed that 'our team', the good guys, were being out-rocketed by the bad guys. Further adding to the dismal time was the fact that Sputnik 2 weighed around 500 kilograms compared to the Vanguard "grapefruit" that weighed a mere 1.4 kg. The Soviets seemed to be way ahead in the space race. What would become of us?

Though unintended, the success of Sputnik 2, as well as the very public televised failure of the American Vanguard launch later in the year, led to the "space race" between the Soviet Union and America, much to the delight of those eager young observers.

By way of explanation, for boys like us the world was indeed divided between the 'good guys', Australians, Americans and the British, and the baddies – the Soviets and the peoples of the Union of Soviet Socialist Republics (USSR). Needless to say the view was encouraged by the press and leaders of the country and our schools. A few years later, in 1962, as Owen had progressed from Middle School to Senior School, the Cold War looked as if it might become seriously hot. The Soviet Premier, Nikita Kruschev, sent missiles to Cuba and they were able to deliver nuclear warheads to the US. Photos from US spy planes clearly showed the missiles on board ships headed for what President Kennedy called "That Imprisoned Island", and clearings in the Cuban jungle where they were to be based. President Kennedy called Kruschev's bluff and, after a few days – weeks it seemed – the missiles were withdrawn. It was kept secret at the time, but Kruschev insisted on what President Trump might call "a deal" that US Jupiter missiles based in Turkey and aimed at the USSR would be removed, and they were a few months later.

Nevertheless, the risk of all-out nuclear war seemed to be high. He clearly recalls the tension and dread that he felt as he regularly looked skyward to see if there was a mushroom cloud. Mercifully, there never was.

There is a downside to all this rocketry and space stuff.

---

sputnik-2-the-space-age-australia-never-had/>. Accessed June 2017.

* * *

Arriving at Orientation Week at the University of Melbourne in February 1965 was a rather frightening moment for Owen as he realised that life beyond the sheltered walls of school had begun. The fear evaporated when he discovered the Melbourne University Astronautical Society (MUAS) stand. Here were a bunch of people who shared our interest in space and things that go there. Owen joined immediately.

* * *

Before beginning the story of MUAS however, we had better understand the times. To coin a huge understatement, they were rather different from today! In all sorts of ways, that is the key to the story.

FASHION DRAMA IN 3 ACTS

● With Light Fingers' Melbourne Cup finish one of the closest ever, racegoers had the added excitement of another Carnival drama. This was in the fashion field — the stars, English model Jean Shrimpton; Paris model Christine Borge. These pictures show the three acts — "The Shrimp" at right; Christine Borge, below.

THE DERBY: Casual shift-style — no stockings, no hat, no gloves, such a short skirt — that shocked Melbourne.

THE CUP: Conservative ensemble that saddened "The Shrimp," shown with Terence Stamp.

THE OAKS: Dress and buttoned jacket that was a happy compromise.

Plate 2.1 *In 1965 English model Jean Shrimpton shocked Melbourne race goers at the Racing Carnival when she wore miniskirts, no stockings, no hat (except at the Melbourne Cup) and no gloves. Such was convention at the time that this was news and a shock to Melbourne society. (Image Credit: Bauer Media Pty Limited / The Australian Women's Weekly November 1965)*

CHAPTER 2

No Boundaries—The Times, They are a' Changing

In this chapter, we will briefly look at the times in which the overseas space industry started and grew. Since then many, many things have changed – obviously technology has developed hugely, but also people's attitudes and government have changed enormously. Nevertheless, the 60s, and for that matter the 70s, were nothing if not exciting times for the young, at least, to live in.

We should not discount the effects of World War II, even well after its conclusion. Although the wartime government tried to keep news of attacks on Darwin and Broome from lowering people's morale, Sydney was targeted by midget submarines that passed the submarine nets protecting the Harbour and attacked shipping. As an aside, Owen's father surveyed the torpedo test range in the Harbour, where torpedoes were test fired to measure their bias before being sent into action. The rationing that so much affected Britain was felt in Australia as well – Owen recalls in the early 50s the Children's Hospital in Melbourne asked for donations of eggs for the children under its care.

The Cold War followed World War II, the Malayan Emergency and the Korean War. The Berlin Wall divided East and West Germany and the Cuban Missile Crisis nearly plunged the world into all-out nuclear war.

In many ways, the defining event (or series of events) of the period was the Vietnam War, or The American War as it is popularly known in Viet Nam, which grew out of France's embarrassing defeat at Dien Ben Phu in Indochina, in what became Communist North Vietnam. Following the Treaty of Paris, the French exited Cambodia, Laos and Vietnam, which was partitioned into North and South Vietnam. France had asked the US to use nuclear weapons against the Viet Cong but, mercifully on advice from the UK, it did not. It is most unlikely (in Owen's opinion) that it would have achieved anything other than universal condemnation and strengthened the determination of the Communist leaders.

The Vietnam War started with US advisers being sent to assist South Vietnam's army in 1955, with ever increasing numbers and eventually large numbers of American and Australian combat troops. For the record, and contrary to popular belief, Australia was asked to help South Vietnam – Owen's father told him that he saw President Diem's letter to the Australian government requesting assistance. (He worked for the military intelligence unit.)

The community's resistance to the war increased throuhout the 1960s, culminating in civil disobedience and huge anti-war marches throughout the US and Australia. Not only was there opposition to the war itself, but also to the conscription of young Australian men based on their date of birth. Vietnam's war grew in intensity as more and more Americans and Australians were thrown into it and more and more were injured and died.

Among the worst decisions that the US military made was to allow and encourage the press to report freely on the war. With hazy black and white television reaching Australia's living rooms at the same time, graphic pictures of death, injury, misery and destruction were relayed directly into people's consciousness and consciences. With conscription and perhaps your neighbour's son being injured in Vietnam, the war was no longer something in a strange faraway land but being played out in the living room and next door. It was not popular, and by 1973 the politicians had decided that it was not worth it and the war ended in ignominy for the US and Australia. Owen records with shame the pictures of Australian citizens publicly condemning soldiers, including conscripts, returning from Vietnam after risking their lives and doing what the government asked of them.

At the same time, the popular culture was developing – Johnny Ray was shocking adults in the 50s and 60s with his music and lyrics while delighting the youth. Owen recalls a concert he gave at the Sidney Myer Music Bowl in the 1950s that Owen couldn't attend although he did hear some of his songs from outside the arena. As popular music developed and expressed the thoughts and feelings of an optimistic generation born during and after the war, but never having really known the horrors of war, we began to question the norms and politics of the day.

It is reasonable to ask whether the war fuelled the development of popular culture or the growing popular culture fuelled the global Peace Movement and its opposition to the Vietnam War. Probably each grew from, and fuelled, the other. Whatever the causes and effects, opposition to the war, popular culture and questioning grew during the 1950s and 1960s. Things were changing and the 60s was an exciting time in Australia. Not just questioning and changing, but also growing in optimism and prosperity. Ordinary people were buying homes with electricity, running water and sewerage, cars, children's bicycles – and televisions.

By the mid 1960s, the Beatles with their funny accents and lyrics that challenged our elders were engaging the youth as never before and encouraging us to further question the state of society. Even world leaders were no longer immune from being protested. When President Johnson visited Melbourne in 1966, huge numbers of students planned to protest Johnson's escalation of the Vietnam War and they gathered along the route of his motorcade shouting, 'LBJ, LBJ, how many kids have you killed today?' A banner on a balcony opposite the University apparently welcomed Johnson but was to be dropped to reveal a protest message as his car passed. The authorities bypassed Melbourne University at the last moment leaving the banner missing its target. Nevertheless, two brothers later managed to throw a paint bomb at him, which missed but hit the unfortunately named Rufus Youngblood, a US secret service agent who thereafter was not very secret.

The peace movement, it seemed, was synonymous with drugs and free love. "Recreational" drugs such as marijuana and LSD became readily available, especially for the young, as did sexual freedom, which came to a grinding halt in the early 1980s with the identification of AIDS and the HIV virus.

Another area where rapid changes were taking place was in clothing. No longer did we have to wear suits or sports jackets when at lectures or outside. "Neat casual" was okay now, even jeans. But women's fashion was hardest hit. Conservative Melbourne was horrified when English model Jean Shrimpton wore a mini-skirt, with no hat, no gloves and no stockings in the Members' Enclosure at the 1965 Melbourne Cup horse race. Heaven forbid, no stockings! The world was about to end … but it didn't. In fact, women were beginning to find their own freedom, not be tied down to home and children and even have their own careers. Germaine Greer's book, the Female Eunuch, appeared in 1970.

In areas that interested us more, cars were getting bigger, more powerful and somehow more strident with their acres of chrome, fins, V8 motors and bench seats. Some English and European cars seemed to move in a more peaceful direction with more rounded lines, such as the Saab and MG. Australian-made Holdens, of course, mirrored the American genre of a big flat bonnet and fins.

Much more important, however, was the march of technology, which built on the discoveries and engineering of the various wars when the military dominated research and technology development. Nuclear weapons and nuclear power spring to mind, but much more important strides were being taken in materials research and development. The obvious examples were transistor radios (but still there were many valve televisions) but behind the scenes the stage was being set for revolutions in materials of all kinds: plastics, steel, aluminium, concrete and even timber. The fundamental technology of the Global Positioning System (GPS) – super accurate clocks – was being investigated.

Computers were growing in power, size and numbers and there was even an IBM mainframe or two in Melbourne. These were huge monsters with flea-like computing power compared even with today's laptops (or even watches), let alone today's super computers. The University of Melbourne's IBM 7044 machine was installed in the early 1960s, the second in Melbourne after IBM Office's own machine. The central processing unit was five or six metres long and two high. Electrical connections between circuit boards were made by wrapping special wires around posts to make those connections. The mat of wiring was as much as 10 cm thick over much of the five metre length of the backplane and two-metre height of the computer. Despite the tens – perhaps hundreds – of thousands of wire-wrapped connections, there was only one failure throughout its life. There were ten or so magnetic tape units and a miserable amount of memory (32,000 words, each 36 bits). Owen wrote his first program for it with punched cards, one card to a statement, in

1965 – or was it 1966. Unsurprisingly, it was a program to predict the direction of a satellite from a point on the earth.

Despite what we think of today as antiquated, the IBM 7040/7044 and 7090 computers were built with transistors which greatly reduced the size, electricity consumption and cooling requirements and so were a considerable step forward in computing over valve computers. There was much prestige in the University having a computer and so it was given its own cavernous room with glass windows for all to see and a considerable staff to tend to its every need. We would later see the next generation of IBM machines being developed, but that's for later in this story.

Jet aircraft in their current form and shape had made their entry but by no means was it commonplace for people to travel beyond capital cities. International travel was ridiculously expensive by today's standards.

In many ways the times were defined in terms of what was not available when compared to today. Television was, of course, analogue, black-and-white and fuzzy. The picture frequently became noisy without an adequate antenna and, in the early days, there were few recorded shows – it was all live broadcast. Certainly nothing from overseas, or even interstate, unless the recording tape was hand-carried to the broadcasting studio. Those who remember Graham Kennedy's In Melbourne Tonight show will continue to claim that it was the best evening entertainment ever produced.

Telephones were connected to their exchange by copper wires. Local calls were fixed price, unlike the UK system where all calls, even local ones, were timed and charged accordingly. Interstate calls in Australia were timed and expensive while international calls were unthinkable. Indeed, the first international call that Owen had known was … well, that will have to wait until we reach that moment in this story. There were no personal computers, no internet, no world wide web, no emails and no social media. Life was simpler then.

On the other hand, President Kennedy had, just four years after Sputnik 1, committed the US to land a man on the moon by the end of the 1960s and return him safely to earth. This was simply unthinkable at the time and many felt impossible. Nevertheless, with the goal in front of them, the US people accomplished a truly remarkable feat in a truly remarkable time frame and landed Neil Armstrong and Edwin Aldrin in July 1969. There were enough space-faring feats of daring almost monthly during the 1960s, as designs and ideas were tested and proved to keep these space nuts thoroughly engaged. But not only did manned space flight advance enormously, so did unmanned civil and military satellites as weights, complexity and capabilities increased.

In some respects though, the biggest difference then was the relative lack of regulation. Today we can hardly move but for a huge mound of regulations, training and certification. A University of Melbourne "cubesat" built, like Australis by students, was faced with a three-quarters-of-a-million

Plate 2.2

dollar fee to get permission from the Australian government to be launched elsewhere! If you wanted to do something in the 60s, largely you simply did it. Somehow common sense ruled, not a vast and unapproachable (and sometimes unobtainable) rule book. There were no boundaries, or at least few boundaries, it seemed. We were free.

Now let's meet some of the men who came to be involved in Australis.

* * *

Plate 2.2 *The defining events of the 1960s were the Vietnam War and the Moratorium Marches when people marched in their hundreds of thousands to end the war. This is one of the iconic images of the period taken in 1970 during the visit of the Vice President of the USA, Spiro Agnew, to Australia. Image courtesy of the National Archives of Australia. NAA: A1200, L85635*

OWEN MACE

Like most Australians, I am the product of immigrants. On my father's side they were from Warwickshire, Gloucestershire and Nottinghamshire in England. My father, Norman, must have been a particularly gifted and driven child. He related stories to me of evenings when the family could not afford coal for heating and he studied under a blanket for warmth and a candle for light. Winning scholarships to secondary school and to Jesus College, Cambridge University, he graduated in about 1930 with First Class Honours in Arts with a Mathematics major – certainly an odd combination it seems today. He was a first-class soccer player and played for the Cambridge team as well as for English county teams and for England in the Netherlands. I have the pocket from his Cambridge jacket signifying that he won a Cambridge Blue for football.

My mother had a rather different background, also a family made good. On her mother's side there were two first fleeters who arrived in 1788 as guests of His Majesty and provided accommodation at His Majesty's Hotel – they were convicts. One is said to have escaped the hangman's noose twice, was sentenced to transportation for life, was Jewish, became Australia's first policeman, was freed, became an innkeeper who defied the rum corps, returned to England and unsuccessfully sued Governor Bourke. Another, aged 19, was transported in 1832 for seven years for forging a promissory note for £200 – a very large sum in the 1830s. Perhaps his rather pathetic note to his mother and sister, along with a bottle of laudanum (used to ease toothaches and commit suicide in those days) caused the judge to impose a very light sentence. He was very lucky!

Another branch, and this is the last convict, was a member of the expedition that attempted to found a settlement at Sorrento on Port Phillip before moving to Van Diemen's Land (Tasmania) and founding Hobart. The next generation moved to the Western District in Victoria where they ran huge sheep stations and were part of the development of Australia's wool industry, which was Australia's largest single export for a century and a half.

I began primary school and then boarding school in Geelong where, in my first year, Sputnik sparked my interest in space. I clipped every newspaper article I could find and was given every book on space that could be found, including my favourite, The Conquest of Space by Chesley Bonestell, which is a collection of imaginative paintings of space scenes, as he imagined them in the 1950s.

While at school, my interests lay with what we today call the STEM subjects (Science, Technology, Engineering and Mathematics) – a rather corny acronym. Actually my subjects were Mathematics, Physics and Chemistry, but there's no decent acronym for that. I hated English as I just couldn't spell and so my essays were returned covered in rivers of red ink by my talented and long-suffering

OVER 2000 OF US SET OFF FROM PADDO

NSW UNI ATTENDS MORATORIUM

SHOOTIN' REVOLUTIONARY THINGS

Plate 2.3

teachers. Nevertheless, something of their lessons must have sunk in because, once spelling and grammar checkers became available on the early computers in 1980, or thereabouts, I could write without the embarrassment of so much red ink and I began to enjoy writing and some of those learnings returned. Yes, I have since personally thanked one of my English teachers for his supreme patience.

* * *

Now let's see what happened to us when we entered tertiary education at the University of Melbourne.

Plate 2.3 *Vietnam Moratorium protesters, Tharunka (Kensington, NSW 1953 -2010) on 14 May 1970. (Image reproduced with the permission of Arc @ UNSW Ltd)*

Plate 3.1 *MUAS Club room; the 'rooftop garret'.*

CHAPTER 3
MUAS—Our Launching Pad

How did this disparate group of people meet? Here is Owen's recollection.

Orientation Week, February 1965, saw the many University clubs jostling with one another to attract the attention of the wide-eyed freshmen, or first-year students as we were then known. There were political clubs with choices between Young Liberal, Labour and, I believe, even Communist clubs; there were sports clubs, perhaps even the lacrosse club that my grandfather played for in 1899 – I have a team photo with him, fresh faced as I undoubtedly was, and the top-hatted Professor Laurie, the club President. There were also dramatic, amateur radio, debating and who knows what else societies. But the one that grabbed my attention was the Melbourne University Astronautical Society (MUAS) advertising that they were interested in satellites and space. That's for me, so I joined.

The Society had a meeting room, or club room, in the rooftops behind a lecture theatre of the "Natural Philosophy" building. These days we'd call it Physics, but never mind: Natural Philosophy it was, housing lecture theatres, student laboratories for practical work, staff offices and laboratories. Somewhere hidden within the building, and I never saw it, was a cyclotron for accelerating electrons to high speeds and energies for research. When operating, the cyclotron played merry hell if we were trying to receive weak satellite signals by drowning them with electronic noise. The cyclotron was sold off decades later and used to produce x-rays for medical diagnosis and treatment.

Plate 3.2

Plate 3.2 *Steve dutifully pointing MUAS' helix antenna on the roof of the Natural Philosophy building. Our 'rooftop garret' is among the roofs out of the picture to the left.*

Richard reached the University two years before Owen. Here is his story.

At last, school was over for me – I was admitted to the University of Melbourne to study Law. Why was an 18-year-old space nut doing Law? As I said, I was hopeless at Maths, Physics and Chemistry – a STEM to me (although the acronym had yet to be coined) was the bit between the root of a plant and the lower branches. A careers person tested me over several hours and concluded that I should be a Librarian or a Lawyer. Law sounded more interesting. In those days, you didn't have to get a Year 12 score in the 90s to get into the Law Faculty – I only achieved a good Pass at Matriculation, as it was then called.

So there I was – Melbourne University, March 1962. Again, as with Owen, I wandered through the Orientation Week displays, and there was MUAS – the Melbourne University Astronautical Society. And there were guys (I don't recall any girls), surrounded by pictures of NASA and Soviet spacecraft, busily trying to

Plate 3.3

Plate 3.3 *Rod and Steve on the roof of the MUAS rooftop garret.*

recruit new members. They didn't have to try at all with me: I fell into their arms. There really were other space nuts like me – I had come home.

Owen has described his introduction to the MUAS, three years later, with the same excitement. When I joined, the Society's main activity was showing space-related films that they had borrowed from the NASA's Australian office in Melbourne. That had been set up because they had a lot of very valuable tracking stations spread around the country, particularly for the manned Mercury, Gemini and the upcoming Apollo programs. We then moved on to the next logical step – tracking satellites, and we built a single, and then a quad helix antenna on the roof of the University's Physics building. In those easy-going days it didn't take much to persuade the Physics Department that a satellite tracking station on their roof was a good idea, and could even enhance their image. The launch by NASA of the TIROS 8 weather satellite in December 1963 allowed us to receive pictures of clouds and the Earth below on a facsimile machine loaned by the Weather Bureau. Then, three years later, NASA's Applications Technology Satellite 1 in stationary orbit, 37,000 kilometres above the Pacific, provided spectacular coverage of a third of the planet.

But to the young space enthusiasts of the MUAS, the ultimate goal, surely, was the building of a satellite. That was how Australis started – one or more of us began talking about making our own satellite. No one had made one in Australia. That didn't matter to us: when people we talked to, "grown ups", told us it couldn't be done, we ignored them. It has been said that human beings in their teens and early 20s appear to be programmed to pursue adventure, to take risks, to try things that their parents would consider dangerous – we had every reason to fail, but we didn't fear failure. In fact, I didn't believe we even thought about it. People would ask, 'But is this project "official"?' To which we replied, 'No – we're just doing it'.

And so Australis was built, and this nerdy law student with the Buddy Holly glasses, although intellectually and practically incapable of designing or constructing any part of the satellite, nevertheless found administration and management tasks to do and contributed what he could.

* * *

Here's Peter's story:

My introduction to space was altogether different. Refusing to have anything to do with wearing glasses, I couldn't see Sputnik 1 when it was launched.

Around the time of the first Sputnik, I was interested in electronics and built electronic instruments, such as a valve oscilloscope, and transistor radios using the earliest semiconductors. I ventured into amateur radio and built my own receivers and transmitters and even managed to get my limited amateur license for the Very High Frequency (VHF) bands as I didn't want to learn Morse Code for the lower frequencies.

These hobbies kept me busy and, of course, I was interested in hearing about the space exploits of the USSR and USA. At University, I did a course in Astronomy. Around this time, computers were starting to appear and this started my interest in computing. After graduating, the first minicomputers started to appear and I built my own version of a Data General Nova computer. It was only through contact with the MUAS that I became interested in satellites.

* * *

The MUAS club room was reached through a lecture theatre, onto a flat roof, then up a steep ladder to enter the cosy 2x3x4 m space with a single window.[1] In it, there was a desk along one wall and several chairs. On the table was a vast piece of equipment coloured battleship grey, for it was indeed a piece of cast-off Royal Australian Navy equipment. This, it was proudly announced, was the B40 and this was our radio receiver for listening to satellites. A capable receiver it was, too. Lots of knobs labelled with strange abbreviations, such as BFO, that we were to learn and understand in time.

Plate 3.4

The B40[2] was built for the Royal Navy and the Royal Australian Navy before it was cast away and found its way to a disposal shop where it was bought for the club. (Disposal shops that sold cast-off military equipment were common in Australian cities in the 1950s and 60s. They had all manner of wondrous equipment that excited amateur radio operators and MUAS members.) The B40s were manufactured in the 1950s by the Murphy Radio Ltd company in Welwyn Garden City in England and each cost the lucky customers £500. This monster was 330 mm wide, 483 mm high, 406 mm deep and weighed 46 kg – no lightweight, but then it didn't need to be on a navy ship or submarine.

1   How to gain the attention of someone in the club room after hours when the building was closed? Answer: a pair of wires from a bell in the club room running covertly down the outside wall to 2 nails. Bridging the nails with a coin rang a bell to alert the occupier to go down and open the main door.

2   Radio Museum, 'Naval Communications Receiver B40' at <http://www.radiomuseum.org/r/murphy_b40_b_40.html>. Accessed June 2017.

Plate 3.4 *The Quad Helix attached to a lift well with Paul working on it.*

The Radio Museum website describes it as being 'NOT Cathedral nor decorative'. You might not agree with the Cathedral part (it was big enough) but you certainly had to agree that it was not decorative. It was, however, functional.

For the technically minded, it was a High Frequency (HF) superheterodyne receiver, 640 kHz to 30.6 MHz in three bands with an unusual intermediate frequency of 500 kHz. Amazingly, it sported 14 thermionic valves, not the thousands of transistors that would be expected in modern receivers.

Unfortunately, however, there were certain limitations to this monster. For a start, we were reliant on a single receiver – there were no spare receivers but the parts most likely to fail, the valves, were available.

It was limited in other ways, too. The frequency coverage was not suitable for receiving signals from satellites and it was limited to amplitude modulation but it did cover the lower frequency amateur radio bands. Club members made "pre-amps" that transformed the favoured satellite frequencies around 136 MHz down into the HF bands reached by the B40.

Entering the room were some black cables that brought signals from the antenna, with pre-amp attached, back to the B40. But where was the antenna? Some 30 or more metres away on a large flat area of roof was our antenna, a clumsy contraption of aluminium pipes and wires standing on a post. It was steered

Plate 3.5

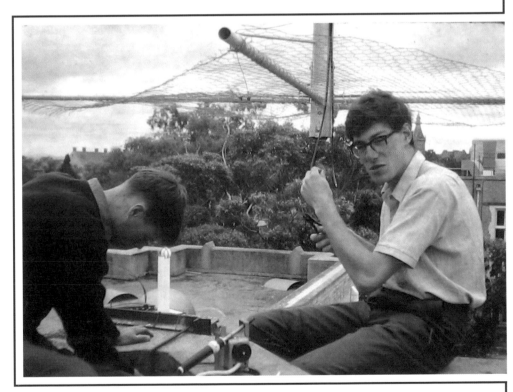

Plates 3.5 *Youthful Paul and Richard working on the Quad Helix antenna.*

Plate 3.6

manually by someone without much of a clue as to where the satellite was, and thus where to point it, and unable to listen to the whistles and beeps (or, more often, the electronic noise) from our B40 receiver. 'Move it a little to the left' shouted across the rooftops to the man hanging onto the antenna handle. Not very satisfactory, so an intercom was installed to give the antenna man half a chance to find the source of our hoped-for satellite. Besides, he could then hear what the B40 was producing and let him decide how to maximise the signal.

It is unfair that the antenna should be called a "contraption" or referred to as Heath Robinson. It was, in fact, a helix antenna with a not inconsiderable gain, and band width and beam width suitable for our uses. Although most satellite transmissions in those days were linearly polarised, unknown satellite attitude relative to the antenna and Faraday rotation in the ionosphere demanded an antenna that was agnostic to the angle of the polarised signal, so a circularly polarised antenna was the solution.

Even so, this was less than ideal. What was really needed was an antenna that could be controlled by the receiver operator in the club room. A design for a "quad helix" antenna consisting of four of our existing antennas working together was needed. There followed a lot of discussion before the experts were satisfied and, needless to say, neither Richard nor Owen contributed anything to the conversation. Richard admits that he contributed even less, except to be mightily impressed at what these tech nerds were doing.

Now this quad helix antenna was no minor construction. Each helix antenna was about three to four metres long and the design called for them to be mounted on a square and separated by about two-and-a-half metres. This meant a device about 4x4x3 metres that had to be pointed at the sky and steered to follow satellites while being entirely remotely controlled from the club room, which was about thirty metres away and not visible from it. An interesting engineering challenge for a group of mostly undergraduate students.

With higher gain, Owen learned, came narrower beam widths. (He was then a first-year engineering student.) This meant that the new antenna had to be pointed more accurately at a satellite if it were to receive its signals and, in turn, a display of the pointing angles of the quad helix was required.

Plate 3.6 *APT pictures on wet fax paper showing the white antarctic coast in the bottom with the Ross Sea partly covered by cloud in the lower right corner. Moving northwards (and upwards), the second picture shows a huge low pressure system and part of the east coast of New Zealand on the right and highlighted by sun glint.*

The design called for an H-frame to which each antenna was attached at the four points of the H. A substantial vertical pole held the H and the helices aloft. In order to point the helices at the satellite, there was a truck bearing at the centre of the H for the elevation axis and the vertical pole rotated in azimuth, also a truck bearing. This was a substantial piece of mechanical engineering, visible from around the University, and it had to withstand the worst that Melbourne weather could throw at it. Azimuth and elevation angles had to be read at the control station beside the B40.

Richard, the obvious leader of the group (lawyers do have some uses), organised weekend working bees to build the quad helix and was perhaps the hardest worker of the group. Nevertheless, we all contributed and, in really quite a short time, the antenna itself was built. Several nagging questions remained – where to mount it, how to drive the azimuth and elevation, and how to measure those 'Az' and 'El' angles.

For the technically minded, the specifications of the Quad Helix were:

Configuration: Four 3.5 m long helix antennas spaced 1.5 wavelengths apart

Centre frequency: 137 MHz

Gain: 22 dB (right hand circular polarisation)

The latter problem was fairly easily solved by the amateur radio operators, or "hams" as they were often called, in the group. They knew that "Synchros" were commonly used by those amateurs to measure antenna azimuth angle and were readily available from the disposal stores. A transmitter synchro on the antenna transmitted angle by way of three alternating current signals and a receiver synchro drove a shaft with pointer attached to display angle. Some experimentation was required to ensure that the three-phase AC signal wires were connected in the correct order so that the pointer followed the antenna in the right direction, but that was easily solved and the pointer calibrated. So that problem was out of the way.

Where to mount the quad helix? Well that was fairly easy, too – there was a lift shaft near to our existing "single" helix antenna, so we could simply attach it to the brick shaft. The Physics Department wouldn't mind, would they? Who knows, they were probably never asked, but they certainly never questioned it. Perhaps it was seen as a status symbol on top of their building. However, it did mean that the vertical, azimuth axis pole had to be well above the top of the lift shaft and so a steel pole was selected (yes, from disposals).[3]

Now for the more difficult problem: how to drive the antenna in azimuth and elevation? In order to explain our solution, dear reader, we must take a short diversion from the narrative.

Somehow, we found that aircraft propeller feathering motors were likely to be powerful enough to

---

3   Paul recalls worrying that, following windy nights, he might arrive the next day to find it draped over the building having caused major damage. He and Owen were relieved when it was finally dismantled.

drive both azimuth and elevation. These motors had an enormous reduction gearbox (9,000 to 1) with correspondingly enormous torque.

We soon found ourselves driving to Melbourne's airport, the Essendon Airport, and Ansett Airline's maintenance department to collect two electric motors from a Convair aircraft. These were heavy beasts, meaning an even stronger vertical pole and a suitably strong mount for the azimuth motor. There must have been considerable trust put in the azimuth drive as it bore the weight of the entire quad helix, something that it was definitely not designed to do. Thank you Lockheed for your excellent design. And thanks to Paul for manufacturing the heavy components which he, Richard, Owen and others assembled into the helix antennas ourselves.

The two feathering motors required a 24 volt supply at substantial current to drive them: barely a challenge for our hams. Underrated relays switched and reversed the voltage so that we could drive the axes in both directions and limit switches warned the operator in the club room of impending disaster. In the clubroom a control panel was built with variable transformers (Variacs) to supply a variable voltage (via diodes, of course) to drive the azimuth and elevation motors.

Consider the further imposition on the building: the mass of cables between the antenna and club room: two cables, one each for the synchros; two heavy cables to carry the drive currents for the motors; four signal wires for the limit switches and intercom. We even had the foresight to include spare wires to replace those frazzled by short circuits.

After considerable effort in 1965 and 1966 and with the staff of the Physics Department ignoring our impositions on the building, our quad helix was ready to be powered for the first time. Variacs slowly turned up and, to our delight and amazement, the quad helix turned and nodded majestically, as appropriate for such a large beast. It worked! There was some argument and rework to "phase" the synchros correctly: it can be somewhat disconcerting for the driver to be shown that the antenna is rotating clockwise when in fact it is anti-clockwise. Nevertheless, these minor problems were overcome quickly and it was operational. Now to listen to some satellites!

* * *

The MUAS group was used to listening to various satellites using the single helix antenna. The Soviets had an active lunar program,[4] even before President Kennedy announced his intention to see an American on the moon. Perhaps it was the Soviets' interest in the moon that led him to issue his bold challenge. Three Soviet spacecraft had been launched towards the moon – Luna-1 (January 1959) was the first spacecraft to escape the Earth's gravity, Luna-2 crashed onto the moon in September of the same year and Luna-3 photographed the far side of the moon in singularly poor

---

4   http://www.Sovietspaceweb.com/spacecraft_planetary_lunar.html

resolution, but it was mankind's first view of that part of our natural satellite and a considerable technological achievement at the time.

In 1965, six lunar shots by the Soviets formed part of their program, to match the US effort. All failed. The following year was rather more successful, with one achieving a soft landing and three orbiting. One of these was announced and tracked by MUAS, though we remain wondering if we heard the Luna or some terrestrial transmission. Whatever the true source of the signal, the press was notified of our intention to listen into the spacecraft's signals and it duly reported our "success". With all the space shots engaging the public's interest, we found a ready audience in the press, radio and television.

Even with "our" satellites, that is American, we knew little if anything about how to decode the signals and so we could not gather any meaningful information from them, so we simply heard squeals and beeps and reported them to an interested press who published whatever we told them. If the beeps and burps happened to be a Soviet spacecraft, so be it.

How satisfying it would be if we could actually make something of the signals and decode, say, spacecraft temperature?

Plate 3.7

Plate 3.7 *The upper APT picture clearly shows the Great Australian Bight (lower centre) and the island of Tasmania (lower right side).*

* * *

Richard recognised that the 1963 announcement of the launch of TIROS 8 (Television Infrared Observation Satellite) satellite[5] might be a satellite that MUAS could really see the data that it was transmitting. For the first time, a satellite was to transmit images of clouds below it in a form that possibly MUAS could receive and actually decode to produce recognisable pictures. Unfortunately, the press announcement did not include any information about how to turn the radio signal into a picture, but the NASA's Senior Scientific Representative in Australia was based in the city of Melbourne and a short tram ride from the University in Swanston Street. A telephone call yielded a meeting with the genial Willson Hunter, who provided us with the technical information we needed.

---

5 NASA, The Television Infrared Observation Program (TIROS)' at <https://science.nasa.gov./missions/tiros>. Accessed June 2017.

The frequency, 137 MHz, was within our capability but the modulation was somewhat inconvenient – the 2.4 kHz subcarrier was frequency modulated (FM) onto the radio frequency (RF) carrier of 137 MHz. However, the faithful B40 could only demodulate amplitude modulated (AM) signals. A FM receiver costing hundreds of dollars was well beyond our means. What to do?

The answer was to dive into the innards of the B40 and pick up the intermediate frequency signal at 500 kHz and build an FM demodulator; a fairly straightforward task for the hams in the group. When Owen arrived at the club there was a vigorous discussion about the success or otherwise of the demodulator. From his unknowing viewpoint, it seemed to work very well, even though a more modern receiver with FM capabilities would have been a bonus. The B40 continued to play its part for years.[6]

The output of the FM demodulator was a 2.4 kHz subcarrier on which the scan lines of the image were amplitude modulated with line and frame synchronisation signals, rather like an analogue television signal, though much slower, of course – two lines of the image per second. There were also bright and dark calibrations. This was a standard format, called Automatic Picture Transmission (APT), for meteorological organisations to send images to each other, typically on land lines, so equipment to display our cloud images was certainly available, but how to get hold of such a "facsimile" machine to capture and record our images?

Plate 3.8

Perhaps the Bureau of Meteorology would be interested in helping us: after all, surely they would be interested in the images we produced. (It was commonly known as the Bureau then, not BOM, as it is now.) Through contacts with the Bureau, the loan of a high-quality Swiss-made Kudelski Nagra tape recorder was arranged and so now we could record the

6   An Eddystone receiver, capable of demodulating FM signals was later found for MUAS.

3.8

Plate 3.8 *Operator's console of the University of Melbourne's IBM 7044 mainframe computer.*

2.4 kHz audio signal, which was then taken to a "fax" machine at the Bureau to be replayed. This at least proved that useful images could be received and provided to Bureau meteorologists, though we understand that there was some resistance to using images of clouds for their forecasts: after all, what could clouds really tell you about the weather? Although not released until 1969, the words of Joni Mitchell's song sums up the initial scepticism of some weather people towards the new technology –

> I've looked at clouds from both sides now
> From up and down and still somehow
> It's clouds illusions I recall
> I really don't know clouds at all.

However, any initial doubts were soon blown away when the first pictures were received. Since then satellites have transformed weather forecasting around the globe.

But back to our role in this exciting new development. Being able to hear, but not see the TIROS 8 data was an unsatisfactory arrangement as there was no immediate confirmation that images were being received, so before long we loaned a fax machine from the Bureau. This did not produce a crisp black-and-white image on a sheet of A4 paper. (In any case, Australia still used the British foolscap paper.) No, the image was burned electrostatically onto a damp, rather fragile, roll of paper (for want of a better word). That is to say, the paper was soaked in a conductive salt that turned dark when an electrical discharge was passed through it.

Imagine it: a machine that drove a pen across a roll of damp paper, discharging sparks to write the image as the roll slowly advanced. Once an image was complete, it was very carefully torn off the roll and then laid out to dry as the next image began to appear.

Despite the complex and obtuse

Plate 3.9

Plate **3.9** *Array of magnetic tape drives of the IBM 7044 computer.*

Plate 3.10

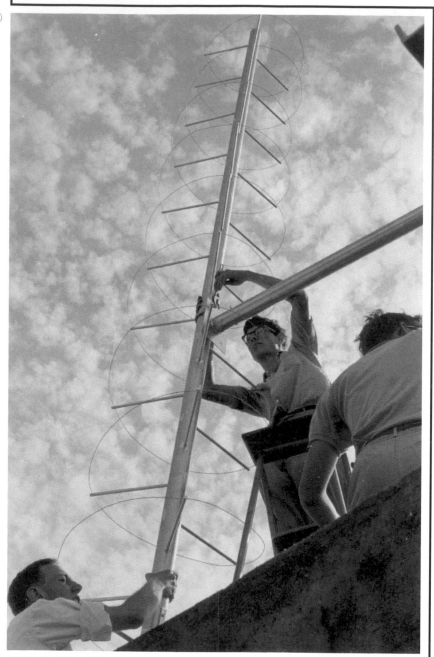

path from RF signal to image, we could see clouds, coastlines and land features. We now had control of the reception of images from the antenna to the image itself.

Satellite APT transmissions continue, although other, much more sophisticated imaging and transmission techniques are used as well. Nowadays, a simple aerial, fairly cheap Very High Frequency (VHF) receiver and personal computer are used by enthusiasts throughout the world to capture images from later generations of APT-transmitting satellites. Furthermore, the images are in colour.

At last, we were doing something useful and real. Richard was employed by the Bureau to rise at some ridiculous time each morning in the dark, motor scooter to the University, let himself into the building, set the equipment going and tape record, on the Nagra, infrared the images from a "pass" of Nimbus 1 and later satellites in the series. Clutching the tape, he then scootered to another university building, hid the tape in a gas meter box and scootered off to bed, for a few hours' sleep before, hopefully,

Plate 3.10 *Building the Quad Helix antenna.*

attending a morning law lecture. Meanwhile, a Commonwealth Government car, with a uniformed Commonwealth Government driver, would arrive at the gas meter box, extract the tape and deliver it to the Bureau for processing. This always seemed to us to be rather cloak-and-dagger stuff, but probably reasonable in the circumstances and the times. It continued until either the Bureau got their own tracking equipment going, or Richard got sick of it – neither of us can remember which.

We, other MUAS members, were happy to contribute to this effort – we were, after all, providing the Bureau with its only regular source of weather satellite pictures at the time. Until something went wrong.

Owen lived in a residential college nearby and early one morning there was a knock on his door. It was Richard explaining that the aerial was not working and there had been a fire. Now Richard, being a lawyer-in-training, was somewhat prone to theatrics – he doesn't mind that being said. 'Okay,' Owen said, 'let's have a look.' Dressing and hurrying to the Natural Philosophy building, he discovered that indeed there had been a fire. That is, a relay or two and some wires had succumbed to the excessive currents imposed upon them. There was nothing that could be done then, so repairs had to wait for daylight and a trip to the electronics shop for more underrated relays and wires – a morning's pictures lost. The relays were simply replaced; after all, they'd worked satisfactorily for months, despite being underrated, and indeed they continued to switch motor direction for years. 'Richard, don't suddenly reverse the motors – wait for a second or two before changing their direction.'

Although images were packed off to the Bureau, the process was to record the 2.4 kHz subcarrier on a professional quality Nagra brand tape recorder, which would then be replayed into the fax machine. This meant the operator could focus on pointing the antenna correctly and coaxing signals from the receiver. During the replay, knobs and dials could be fiddled with to optimise the image quality, such as it was, and any errors could be corrected by simply replaying the tape again. This meant that multiple copies of the weather pictures could be made.

As we studied the images, we began to learn about the different types of clouds and what they meant. We could see cold fronts, a cyclone or two and the "speckled" cloud of cold air behind a front. We were able to anticipate the evening's forecasts from the Bureau. The coastline of Australia, the Great Australian Bight, the Gulf of Carpentaria, the east and west coasts, New Zealand and even Port Phillip Bay were just as the maps had predicted! Occasionally images stretched as far as the unfamiliar Antarctic coast. What a success and what a treat for us techies.

One morning Richard excitedly reported that the first image of a pass that morning was totally blank. This was something that had not happened before. What had gone wrong with our equipment? He replayed the tape – no, just a steady subcarrier with no information on it. 'Wait, play it again, weren't there slight variations? Could that be an image? Unlikely, but let's see if we can get an image onto the fax.' After a number of replays and much knob twiddling, a very low contrast image appeared on the sodden fax paper. It was clear enough, however, to see a very unfamiliar coast with fjords and was it snow and glaciers? Surely not! Well, it was indeed the last image taken by NASA's new experimental weather satellite, Nimbus 1, in the northern hemisphere before it entered the dark side of the Earth and ceased transmitting images. It was part of Greenland!

There had been a fault on board: the flash lamp had failed to erase the image from the detector and the image had slowly faded until the transmission sequence began again over Australia with the read-out of the image to l'il ole MUAS! Willson Hunter was contacted to find out what was going on. After sending tapes and images through him to the Nimbus project office, it was announced that we had seen the first convulsions of the dying satellite. Congratulations, Richard, for persisting with the difficult task of extracting the image from the very weak signal.

* * *

Something that came out of this project was to be of great significance and import to us, though we didn't recognise it at the time. The NASA representative, Willson Hunter, became a friend of MUAS and to Owen and others personally.

As a young engineer in 1945, Willson was in charge of the Icing research in the Wind Tunnels and Flight Division of the National Advisory Committee for Aeronautics (NACA)[7] at the Lewis Laboratory in Cleveland Ohio. He also worked for B. F. Goodrich and Co. researching icing problems on aircraft.

Willson rose through the ranks and was a senior manager in the 1960s, prior to being appointed as NASA's representative in Australia. He was responsible for all NASA's tracking stations and operations in Australia and played critical parts in the successful reception of signals from the manned Apollo flights at NASA's Honeysuckle Creek near Canberra and the Australian radio telescope at Parkes, New South Wales. There is a brief shot of Willson in a 1969 newsreel on Apollo 11, early in the movie 'The Dish'.

Willson's full contribution to AO5 in encouraging us, opening doors and greasing the wheels in NASA will probably never be fully known. Genial and urbane, he and his wife, Marj, were generous and encouraging friends. It was sad indeed to hear of his later demise.

* * *

This is all very well, pointing the quad helix and receiving weather pictures, but where to point the antenna? The satellites' orbits were known precisely and they appeared in two windows each day: several northbound orbits in the morning and several southbound twelve hours later. This is known as a sun synchronous orbit, having the advantage that illumination on the ground remains constant

---

7   NACA was the precursor to and later became the National Aeronautics and Space Administration, NASA.

at any latitude, except for seasonal variations. This makes short-term comparisons easier as well as provides images at the same time each day for the weather forecasters. Nevertheless, a satellite "pass" over a receiving station varies from day to day within each window, and so some form of prediction service is required. When and where to point the antenna – time, azimuth and elevation.

An interesting problem, how to predict the azimuth and elevation of a satellite. We knew that precise orbit parameters were published regularly by the US for their, and some Soviet, satellites. Perhaps that might help. Yes, the time and longitude of an ascending node was included. (The ascending node is the point in a satellite's orbit when it crosses the equator travelling northwards.) Owen explains. After a discussion with his father – and he had been a surveyor – he pointed me at an old surveying book of his. I still have it. Father had bought *The Text Book of Topographical and Geographical Surveying*, published by His Majesty's Stationery Office in London, from Dymock's Book Arcade in Sydney on 25 January 1943 while on leave (and escape) from Sarawak in the middle of World War II.

The Text Book is pure Victorian England, having been edited by Colonel Sir Charles Close K.B.E, C.B., C.M.G.,[8] and a Fellow of the Royal Geographical Society, as Dad was then. Close's co-editor was also a colonel, H. St. J. L. Winterbottom, C.M.G., D.S.O. I imagine them to be pith helmeted and heavily moustachioed, but no doubt that's quite unfair. Wikipedia's account of Close's life shows him as having done survey work in Burma, Nigeria, the Cameroons and throughout Africa, including time as a commander of a surveying unit during the Boer War.

The text is a treasure house of minute details of survey instructions, methods, descriptions and formulas. There is all manner of useful information for surveyors, such as camp equipment, descriptions of surveying and analysis methods, tables and maps of the colonies where Close had worked. 'He was knighted in recognition of the Ordnance Survey's work during WW I when over 30 million maps were produced' according to Wikipedia. Close was Director General of the Ordnance Survey throughout the first World War.

Importantly, *The Text Book* had a section on 'Useful Trigonometrical Formulæ' (note the diphthong, 'æ'), including equations for plane and spherical trigonometry. Immediately, once Owen saw those equations, the light bulb went on: knowing an ascending node, the altitude and period of the nearly circular orbit, he could calculate the location of the sub-satellite point at any moment and hence the azimuth and elevation of it from any point on the Earth. This was the way to provide antenna pointing information for the TIROS passes to an accuracy sufficient for the beam width of the quad helix.

In principle, anyway. The calculations with pencil, paper and slide rule would be impossibly tedious to produce for several passes each day. No, there had to be a better way and there was – the University's brand spanking new IBM 7044 computer! It seems unbelievable these days and yet the

8   Wikipedi, 'Charles Close' at <https://en.wikipedia.org/wiki/Charles_Close>. Accessed June 2017.

IBM 7044 had 64,000 words of memory, or 192 kbytes of magnetic core memory – remember that the very first IBM personal computers sported over three times this memory, 640 kilobytes, and now many gigabytes are the norm.

Minor difficulty: Owen had neither the knowledge to program it nor the permission to use it. No problem – these were days with 'no boundaries' – and permission was soon granted. A course in Fortran programming, some mathematics, coding sheets and much time spent typing on the punched card machine soon produced a printout from the computer showing how many coding errors he had made. Correct the errors, re-punch the cards, submit the small deck of cards again on the next day, for there was only one "run" per day, for the next batch of errors. Eventually, the errors were eliminated and the program ran, producing antenna pointing information, but were the numbers correct? The next morning's passes proved that they were. We now had the means to predict when passes would occur and where to point the antenna. Furthermore, the accuracy of the process was accurate enough for a week or two's passes. Problem solved.

Owen's recollection is that this was in 1966 when he was in his second year of Engineering studies at the University of Melbourne but this does not align with the memories of others, nor with the dates for Tiros 8. Perhaps it was 1965. In any event, it was the start of a lifetime of engagement with computer hardware and software and even at the age of 70+ in retirement, he programs for his own satisfaction and recreation. But not Fortran, which then was a beast of a language!

In any event, with this introduction to computing satellite predictions, Owen was well placed for the next phase of the MUAS story.

* * *

For reasons we can no longer fathom, except perhaps for self-aggrandisement, we publicised our successes in the press, radio and television. This also turned out to be important for the next steps.

* * *

One day at our "formal" weekly meeting discussing satellites, antennas, space, electronics and such like, somebody made the observation that we were successfully tracking satellites and obtaining weather pictures, so why not build one ourselves? No one has yet owned up to being that person. Well … nearly no one. Then the discussion went along the lines of, well, why not, it would be fun, despite there being just twenty or so members of MUAS, some of whom like your authors were not able to contribute much technically.

The innocence of youth, especially in those times. In an article in the British *Classic and Sports Car* magazine about Adrian Squire,[9] founder of the short-lived Squire Car Manufacturing Company in

---

9   Squire was unable to manufacture more than a few of his beautiful sports car and the company folded in 1936. He was killed in an air raid in 1940 at the Bristol Aircraft Company factory at Filton, Bristol where he was working.

Henley-on-Thames, Oxfordshire, the author, observed:

> There is no energy source in the universe quite so powerful and intoxicating as youth. As people age, they tend to compromise, to settle, to reach no longer for the stars but for their spectacles – and then only to see restrictions and obstacles wherever they look. They rationalise. Youthful vigour (or arrogance, depending on your age) recognises no such bounds nor commonsense limitations. Nothing is impossible.

* * *

We will now meet some of those involved with Australis and read the stories of how they came to be involved. Later we will see how they were involved.

* * *

ROD TUCKER

Here's how Rod found the Melbourne University Astronautical Society:

'One cold evening in October 1957, I watched Sputnik cut across the Melbourne sky. I was hooked. Later, in the early 1960s, as a ham radio operator, I was inspired by the involvement of many hams around the world in tracking and sending signals to the early OSCAR amateur radio satellites launched in the US. In 1966, my first year as a student at the University of Melbourne, I visited the Astronautical Society club rooms and watched as weather satellite pictures appeared on an early equivalent of a fax machine. There was a lot of activity in the Astronautical Society at the time, with the construction of AO-5 already underway. I quickly joined the Society and offered to assist with AO-5.'

* * *

PETER HAMMER

Peter recalls his childhood years and later at the University of Melbourne and how he became involved with Australis:

'My parents were born in Berlin in Germany: my father in 1900 and mother about ten years later. They endured the horrors of World War 1 and as a result my father could never eat pumpkin as that was the only vegetable available. He was a businessman and my mother did a PhD in Chemistry. They both had a Jewish background but never set foot in a synagogue. Nevertheless, my father found himself in a concentration camp in the late 1930s. Luckily his brother-in-law was Wilfred Burchett who secured his release and both my parents went to the United Kingdom, destitute.

'Wilfred then went on to get a number of other people out of Germany. He was a humanitarian and later in life ran foul of the Australian Government with his stance against the Vietnam War and lost his passport as a result. Wilfred had a brother whose child is a very well-known Australian chef.

'My parents emigrated to Australia under the 10 pound plan; "ten pound Poms", as they were rather disparagingly called. After arriving in Australia and a short stay in a camp, they established a business manufacturing adhesives. While my father ran the company, my mother produced all the manufacturing recipes while bringing up two boys.

'My mother had been interested in photography for a long time in Germany. As well as taking photographs, she was particularly interested in the technical side of developing and enlarging. Her interest rubbed off on me and by the time the Olympic Games came to Melbourne in 1956, I was busy taking photos. My father was interested in psychology and he was an expert at graphology who assisted the police in a number of cases.'

Peter says that he had a happy childhood and did very well at his private school but his abilities fell away in sport. He left school at the end of Year 11 and enrolled in a Diploma of Applied Physics at the Royal Melbourne Institute of Technology: he was too young to enter University as there was then an age limit.

'I first became interested in electronics when I was doing a Diploma of Applied Physics at the Royal Melbourne Institute of Technology (now RMIT University). After graduating I went straight into the second year of a Physics degree at Melbourne University. Here, I started getting interested in amateur radio and it wasn't long before I had my own call sign VK3ZPI. This was a limited licence restricted to the VHF and higher frequencies. I started building my own transmitters and receivers and often communicated on mobile amateur radio while driving around Melbourne.

'I gradually lost interest in Physics and switched to a Science degree, with an Electronics major. After graduating, I enrolled in a PhD program at the Physics RAAF Department, whose Professor was Vic Hopper. The undergraduate courses were run for air force cadets at the RAAF Academy, Point Cook, whereas the post-graduate school was attached to the University of Melbourne.

'My chosen field was Upper Atmosphere Physics which is the study of the Earth's ionosphere starting at about 75km altitude. My PhD was mainly concerned with developing a new way of studying this region and it meant designing and building specialised transmitters and receivers. Altogether, this took several years of work. The resulting data was recorded on magnetic tape. In those days, the University had just acquired its IBM 7044 computer. As postgraduate students we were allowed to book blocks of time of a few hours a day, running the machine after the daily 9-5 daytime shift. This saw me often running the computer late at night and having to cope with issues when things went wrong, like the line printer spewing out fanfold paper at rates of several metres/ second or the card reader not wanting to work. Luckily there were backup systems.

'The head of my department was very much against spending time on anything but research so there were occasional meetings suggesting I stop all this "silly satellite stuff". To be fair, he was extremely dedicated to research and set an excellent example of single-minded focus on research. Needless to say, I still persevered with my PhD while building bits of satellite and playing with amateur radio. I also ran the ire of the senior electronic technician in the department, nicknamed "squirrel" for his way of locking everything up and making it hard for people designing and building electronics. (It didn't help that his nickname rhymed with his name!)'

* * *

LANCE GINNER

Lance remembers his pivotal role in the creation of the first Amateur Radio Satellites.

'In April 1959, 15 months after the successful launch of America's first satellite, Explorer 1, Don Stoner, amateur radio callsign W6NTS, wrote an article in the leading amateur radio magazine, CQ, suggesting (and indeed, some thought it was a bit of a jest) that amateurs should build their own communications satellite. Remember that the first commercial active communications spacecraft, Telstar 1, was not launched until July 1962, over 3 years later.

'In October 1959 American amateurs Fred Hicks, W6EJU and Chuck Towns, K6LFH contacted Don Stoner and this began an intense process determining how to proceed. Project OSCAR (Orbiting Satellite Carrying Amateur Radio) was formed shortly thereafter, with Chuck Towns, Bernie Barrick W6OON, Stan Benson K6CBK and Nick Marshall attending the first meeting. A "White Paper" was prepared, outlining the objectives of the program – then the hard work really began.

'George Jacobs, W3ASK, volunteered to be the project's Washington contact and he was instrumental, along with OSCAR Board members, in gaining the necessary approvals from the various agencies. By November 1961, a month before the launch of OSCAR 1, approvals had been obtained from the Federal Communications Commission, the State Department, the Undersecretary of the Air Force and the American Radio Relay League. (These guys had very good connections.)

'The approval cycle was made more difficult because the Discoverer spacecraft that OSCAR 1 was going to ride into orbit which was the cover for the classified Corona camera and film return program. I got personally involved with Project OSCAR shortly after its incorporation. I had joined Lockheed in 1960 and was a test conductor for systems checkout of the Discoverer satellites and payload interfaces. I was therefore in a unique position to assist in the integration of OSCAR 1 to the host vehicle, the Agena second stage of the Thor Agena launcher. I was also responsible for integrating and for the assembly and wiring of the electronics that went into OSCARS 1 and 2.'

Richard takes up the story of OSCARS 1 and 2 –

OSCAR 1 was launched with Discoverer 36 (Corona 9029) on 12 December 1961 from Vandenberg Air Force Base, California – a little over 2½ years after Don Stoner had first proposed an amateur radio satellite. Weighing 10 kilograms, it entered a 245 by 474 kilometre polar orbit. OSCAR 1 had a battery powered 140 milliwatt transmitter operating in the amateur 2 metre band. The transmitter worked for 21 days, until the battery was exhausted. It was, like Sputnik 1, a simple satellite, sending out the radio amateur's HI greeting in morse code, the speed of the message being controlled by a temperature sensor inside the craft. It was an historic first, prompting Vice President Johnson to send the OSCAR team a congratulatory telegram –

> 'For me this project is symbolic of the type of freedom for which this country stands – freedom of enterprise and freedom of participation on the part of individuals throughout the world.'

Among other firsts, OSCAR 1 was the world's first piggyback satellite and the first private, non-government spacecraft. It's job done, it re-entered the atmosphere on 31 January 1962.

OSCAR 2 rode into orbit 6 months later, on 2 June 1962, again on a Thor Agena from Vandenberg, with Corona 9036. Nearly identical to OSCAR 1, the thermal coating was changed to achieve a cooler internal temperature and the transmitter power lowered to 100 milliwatts to extend battery life. However, the orbit was lower than OSCAR 1 (394 by 207 kilometres) and OSCAR 2 re-entered the atmosphere after 22 days but it, like its predecessor, gave radio amateurs around the world a taste of what was possible by flying their hobby in space.

Lance continues his story –

'Early design work started on OSCAR 3 in late 1963. Don Norgaard, W6VMH and Ed Hilton, W6VKP, provided the basic 2 metre in-band linear translator design for what was to be a real communications satellite, while I was responsible for the mechanical design and integration of the satellite with the host vehicle. That host was a highly classified program, separate from the Discoverer/Corona series. I happened to be the only one in the OSCAR program who had the necessary security clearances and as such I was limited in what could be said about a launch opportunity. I provided the host vehicle contractor with an outline drawing which showed the satellite envelope and ejection system. I also supplied a baseplate test fixture for them to fit check.

'I personally delivered OSCAR 3 to the Vandenberg launch site and assisted in the mounting of the spacecraft to the host vehicle. Keep in mind this is the very first time that the host customer had ever seen the OSCAR spacecraft – there was no official "certification process." Doing any coordination with the amateur community was difficult due to the security constraints but we did manage to have several "test runs" of our world wide communications network, with one of them "coincidentally" scheduled on the March 1965 launch date – this of course was all pre-Internet.'

Back to Richard about OSCAR 3 –

This satellite was solar powered to extend its life, with a 2 metre to 2 metre translator. Heavier than its two predecessors at 16.3 kilograms, OSCAR 3 was successfully launched from Vandenberg with 7 Air Force satellites into a near polar orbit of 917 by 884 kilometres by the trusty Thor Agena on 9 March 1965. The translator operated successfully for 18 days and the 2 beacon transmitters for several months. More than 1000 amateurs in 22 countries communicated with each other through OSCAR 3. Because of its higher orbit, it remains in space today, circling the earth every 103 minutes.

Lance describes the bold experiment that was OSCAR 4 –

'OSCAR 4 was a very unique case. During the final ground testing of OSCAR 3 we received an invitation to ride on the just being certified Titan 3C rocket that was going to a slow drift synchronous orbit, some 33,000 kilometres high, so that it would gradually move around the earth beneath it. Due to the size of the host vehicle there was no particular weight or size limit. The main issue was, were we interested and could we meet a launch date 12 months away? This was an amazing opportunity and the TRW (Thompson Ramo Wooldridge) Radio Club decided to take up the challenge. Project OSCAR supported the integration of the satellite. OSCAR 4, with a 144 MHz to 432 MHz three watt linear transponder, was launched from Cape Canaveral on 21 December 1965, only nine months after OSCAR 3. The final stage of the Titan did not fire and the orbit achieved was 33,500 kilometres by 162 kilometres, each orbit taking nearly 10 hours. OSCAR 4 turned out to be more successful than many of the other ride along payloads in that we did not depend on highly directional satellite antennas and had world wide reception capability that provided early tracking and several successful two way communications were made, including the first direct satellite link between the United States and the USSR. Oscar 4 operated for 85 days and re entered the atmosphere on 12 April 1976.'

* * *

Jan King

Jan was born in Pontiac, Michigan which was a General Motors town. As he remembers, 'I was in fifth grade the day Sputnik was launched. I was terribly interested in radios and later became a radio amateur. The President (Dwight 'Ike' Eisenhower) came on the air very quickly after the event and said that we have to win this race and I'm calling on all young people to become much more involved in space and in math and science. We need an aerospace community in the United States and this is a call for all those who might be interested to consider that in your career choice.'

The timing was perfect for Jan. He had been involved, with a friend Bill, with radio amateurs building a satellite from his eleventh grade. General Motors offered his school time on their IBM 1620 computer and so Jan and Bill wrote a computer program in Fortran to predict when and where to point their antenna.

He and his mate, Bill, built a large antenna, not unlike the quad helix, on the roof of his home and tracked OSCAR 3 with it.

Jan remembers the early times, when he was in University and first visited California. During that trip, Jan and his lifelong best friend Bill Mazur, visited Project Oscar (Chuck Towns and Lance Ginner). Most significant was the visit to Lance Ginner's garage and the trip to Project Oscar's Lab and Tracking Facility at Foothill's College in Los Altos Hills. We were shown the Australis spacecraft by Lance and it was the first satellite of any kind I had ever seen - for real. I'd seen countless photos but, this satellite actually ended up in orbit and little did I know at the time that I would be fundamentally responsible for getting it there. This would have been in 1967.

At 21, Jan graduated from University and was hired by NASA and assigned to the Goddard Spaceflight Center which is outside of Washington, DC. Here, he was part of a team testing NASA satellites of all types and sizes up to huge astronomical observatories. As Jan says, 'I was like a kid in a candy store seeing all these weird scientific satellites every day ... I had the bug.'

One day, a group of people asked him to join them building an amateur radio satellite. These guys were radio amateurs and had considerable technical and managerial seniority within NASA. They knew of Project OSCAR and that it was having difficulty getting agreement from the US Air Force for further launches.

* * *

THE VW

While a car is not quite the same thing as the people, this one figures throughout this story and so deserves a mention. Owen was exceptionally lucky – he owned a car, a VW beetle , to be precise. It was probably one of the new 1966 models with larger windows and (wait for it) a larger engine – 1,200 cc. For those who have not had the misfortune to ride in one, he simply says that the VW Käfer (German for "beetle") was Adolf Hitler's idea for a people's car. Adolf rode in Mercedes Benz cars.

Rear engine, two-door, it was fully air conditioned by opening the windows, had a reserve fuel tank that was operated by an out-of-reach lever and its rear wheels could fold under the car in certain situations. Nevertheless, it served Australis well and was the vehicle that we used to collect the Ansett prop pitch motors from Essendon airport.

Plate 4.1 *Peter's Very High Frequency (VHF) transmitter on the left and his High Frquency (HF) Transmitter on the right. They are sitting on computer printer fanfolder paper and hollerith cards on which is punched one line of a computer program.*

CHAPTER 4
Australis Electronics

We did not know enough to know that such a hare-brained idea as actually building and launching a satellite could never succeed, so we planned how we might go about it. We quickly realised that without a launch into space, building a satellite was pointless. Australia had not demonstrated a satellite launching capability at that stage and the only countries that had launched satellites were the Soviet Union and America. Obviously, the "baddies" were not a consideration, even if we could speak their language, so an American launch was the only feasible option. Furthermore, our ham members knew that the California-based Orbiting Satellite Carrying Amateur Radio (OSCAR) organisation had already flown three satellites.

The Project OSCAR Inc. was formed in 1960 by amateur radio operators in California who were members of the Lockheed Radio Club of Redondo Beach.[1] (Lockheed was an electronics company specialising in defence electronics.) Many worked at Lockheed, other defence companies and the Foothills College. Many were also members of the teams that built and flew the Discoverer, or Corona, missions. These were highly classified spy satellites that photographed the ground beneath them on 70 mm film which was returned to Earth in a return capsule. The return capsule was supposed to be caught by a cargo plane trailing a rope in the hope of snagging the parachute risers. Unsurprisingly, the capture process was extremely difficult and not initially very successful. The return capsule was also designed to remain afloat for two days to allow recovery from the sea by the US Navy or anyone else passing by. There were 144 Corona satellites launched of which 102 returned usable photographs.

OSCAR 1 was launched on 12 December 1961. It was no coincidence that many of the OSCAR people worked on the Corona programme. In any event, space was found for a small payload on the Agena second stage from which it was ejected, presumably so that it would not interfere with the Corona satellite. It carried a 140 milliwatt (140 mW) transmitter in the two-metre amateur radio band at 144.983 megahertz (144.983 MHz) and transmitted the hams' greeting, HI, in Morse code[2] (four dots followed by two dots) that was received by well over 500 amateurs in 28 countries over the three weeks that the batteries lasted. This was a simple, low-cost satellite – the ejection spring came from a local store and cost less that two US dollars. This was just over four years after the first Sputnik.

---

1   Although TRW is mentioned in the Wikipedia article, we are informed by Lance that 'TRW had no involvement until Oscar 4'. It was the Lockheed Radio Club.

2   Morse Code was devised by Samuel Morse in about 1837 for transmitting characters using a code consisting of dots and dashes, thereby building up text. For radio transmission, the dots and dashes are represented by pulsing a transmitter, or audio tones, on for short or long periods. From Marconi's first radio transmissions in the 1890s, Morse Code was used for long distance radio communication. By the year 2000 it had all but disappeared in favour of other means of communicating text.

This flight was quite an amazing achievement, not just because it was the first satellite to be ejected from another or that it was the first amateur radio satellite. How on Earth (literally) did Project OSCAR persuade the spy agency (the Central Intelligence Agency) and the US Air Force to fly a civilian satellite? The naysayers no doubt would have argued that it might compromise the success of the mission and its security. On the other hand, it helped that some of the proponents worked on the Corona project. In all likelihood senior managers were amateurs also and this would have helped as well.

There is also another aspect. Amateur radio was a far more popular pastime and was more visible than it is today. While today amateurs are frequently the first to make contact from a disaster zone (such as Cyclone Tracy[3] in Darwin on Christmas Day 1974 and more recent cyclones and extreme weather events), their work is rarely acknowledged. Of course, in the 60s, amateurs normally made their own receivers and valve transmitters as well as using them to communicate. As the decades passed, however, commercial receivers and later transmitters became available to amateurs and fewer amateurs built their own equipment. With the internet, computers, mobile phones and tablets, the ease of communication reduced the allure of talking to someone on the other side of the globe on a noisy single sideband channel.[4]

Nevertheless, in the 60s, there were large numbers of amateur radio operators in the US and many in senior technical and management positions.

Plate 4.2

3   Wikipedia, 'Cyclone Tracy' at <https://en.wikipedia.org/wiki/Cyclone_Tracy>. Accessed June 2017.

4   Single sideband is a modulation technique that reduces the amount of bandwidth needed compared with amplitude modulation and so improves the received signal strength. The disadvantage is that, unless the receiver is tuned to the transmit frequency exactly, speech is distorted.

Plate 4.2 *The VHF transmitter during construction. Note the ferrite tuning coils.*

OSCAR 1 was a pioneer in many ways.[5] The use of HI as identification, two-metre signals and the practice of ejecting secondary satellites with springs became something of a standard.

OSCAR 2, launched on 2 June 1962 on another Corona mission as a secondary payload or, less politely, as ballast. It was similar to OSCAR 1 except that it was a rectangular shape at 300x250x120 mm, weighing 10 kg with a single monopole antenna. The transmitter power was lowered to extend battery life and a measurement of its internal temperature was transmitted. Its external surface was changed to lower the internal temperature compared to the previous OSCAR. It transmitted for 18 days until the battery was exhausted.

OSCAR 3, weighing in at 16.3 kg, was launched on 9 March 1965 and carried a very different payload, a linear transponder that relayed signals to extend the range of two-metre communications over thousands of miles. Over 1,000 amateurs communicated using the transponder and received the beacon signals.

OSCAR 4 was planned for a geostationary orbit when launched on 12 December 1965 with a payload similar to its predecessor, but it reached only a geostationary transfer orbit of 33,000km apogee (the highest point in the orbit) and 161km (lowest point). Its linear transponder received signals on 144 MHz and re-transmitted them on 432 MHz. The transmit power was three watts and it carried the first direct satellite communication between the US and the USSR before it re-entered the atmosphere on 16 March 1966.

The Project OSCAR people were visionaries who set their sights high and achieved a number of firsts. Amazing achievements less than ten years after Sputnik – and amateurs at that!

* * *

Well, we thought, if they can do it, so can we. In fact, perhaps the OSCAR people could organise a launch for us? The address of OSCAR headquarters was found – no internet then, remember – and a letter dispatched asking them if we built a satellite, would they organise a launch for us. A reply arrived barely two weeks later. Given that a week was the delivery time for an international letter then, our request must have been replied to immediately. The answer was yes, they would try to launch it if we built it. The challenge had been laid! (And don't worry about the try to launch it part.)

We now had to think carefully about what we wanted to fly. Clearly, it had to be a satellite that would interest amateur radio operators and, besides, some of our members were amateurs themselves. We made contact with the Wireless Institute of Australia, the WIA, who, with their members, supported us magnificently throughout the project.

---

5   Wikipedia, 'OSCAR 1' at <https://en.wikipedia.org/wiki/OSCAR_1>. Accessed June2017, See similarly for the other OSCAR satellites. Also: Andres Bilsong, OSCAR-1 launched 50 Years Ago (online). At <http://www.arrl.org/files/file/Technology/Bilsing.pdf>. Accessed June 2017.

A name was found – Australis - to signify the southern hemisphere and relate to Australia. Project Australis was born with its headquarters in the rooftop garret of the Natural Philosophy building in the University of Melbourne, postal address care of the Student's Union and telephone number ... well, who remembers these days?

So in general terms, there must be something for amateurs in the Australis satellite. But what? We knew, of course, of the satellites that Project OSCAR had flown. The ultimate for an amateur satellite, the "gold standard", therefore would be a transponder that received amateur transmissions on one frequency in the two-metre amateur band, amplified it and re-transmitted on another frequency in the same amateur band. In this way, a two-metre radio transmission, normally limited to not much more than line of sight, can be beamed to a satellite transponder that is hundreds, even a thousand, kilometres away. The signal is then re-transmitted to the receiving station a further thousand kilometres away. The two stations might even be on opposite sides of an ocean.[6]

Technically, a two-metre transponder is a much harder device to design and build than a two metre transmitter. The reason for the technical difficulty of a transponder is that the transmitter would be in close physical and frequency proximity to the receiver. Keeping the receiver deaf to the nearby transmitted signals while at the same time being highly sensitive to its own receiving frequency is a difficult task further complicated by the harsh space environment that was largely unknown to us.[7]

The OSCAR people, whose profession was flying satellites, well understood the space environment and how to build electronics to operate in space. So it was decided to "cut our teeth" on a simple satellite

Plate 4.3

6   This is a standard technique for marine communications which are on frequencies close to the two-metre amateur band. A transponder may be sited on a high headland or hill to extend the range of marine communications for many tens of kilometres or more when line of sight communication for a small recreational vessel might be just ten kilometres.

7   Modern marine transponders accomplish this but the ironmonger required to adequately filter the transmission signal from the receiver was far too large for Australis.

Plate 4.3 *A prototype HF transmitter showing the tuning capacitors mounted on the vertical panel.*

but with the view to future satellites carrying transponders, the ultimate amateur radio satellite. While seemingly a simple decision, it took some discussion to reach it.

To be useful, any future transponder satellite that we dreamed of had to be stabilised in orbit: that is, it must not tumble in orbit, otherwise signals would fade as the satellite's transmission pattern became unfavourable for the amateur's ground transmitting station or the receiving station. This had been a problem with the earlier OSCAR transponder satellites. Therefore, a stabilisation system was required and also a means of measuring the satellite's orientation was needed.

As Australis was to be a demonstration or test satellite showing the way for future amateur radio satellites, we decided to fly two transmitters, a two-metre transmitter as on previous OSCARs and a ten-metre transmitter in the amateur band, a first for any satellite at the time of our design. The transmitters must not just say "HI", they must carry useful information such as battery voltage and current drain. If the ten-metre (High Frequency, HF) transmitter was to be useful, it had to transmit at a considerably higher power than the 100 mW two-metre transmitter in order to overcome the greater path loss at ten metres. In fact, the ten-metre transmitter nominal power was 250 mW – high power indeed for a battery-powered satellite of the size we envisaged. This led to the idea of a command system that could control the "high power" HF transmitter.

Add a battery, a means of maintaining the temperature of the electronics, and separation system, a structure and we have the 15 sub-systems of Australis, which were:

1. A stabilisation system to reduce any residual tumbling in orbit

2. An orientation system to ensure that Australis' antennas are in a direction favourable for receiving the signals on the ground with no, or minimal fading

3. A two-metre, 50 mW transmitter, amplitude modulated by the telemetry signal

4. A ten-metre, 250 mW transmitter, also amplitude modulated by the telemetry signal

5. A command system consisting of a receiver and decoder to turn the ten-metre transmitter on and off

6. Antennas for the two transmitters and the command receiver

7. A "HI keyer" to generate HI in Morse Code

8. A telemetry system to generate the modulation signal for the two transmitters

9. A battery to power the electronics

10. A separation system to eject Australis from the launch rocket, including a master on/off switch

11. A structure, including thermal control. Thermal control was a matter outside our control and finally designed by NASA. Nevertheless, provision had to be made for a temperature control layer to be added to Australis' outer surface.

Without amateurs listening to Australis and reporting the signals they received, the project was worthless, except perhaps as an ego trip for us students. Accordingly, a ground segment was recognised as consisting of:

12. A network of amateurs to receive and report the signals they heard

13. A means of generating and distributing predictions on when and where to look for Australis

14. Command stations to send command signals to control the HF transmitter

15. An analysis centre to analyse reception reports from receiving stations.

Although we did not express the design decisions in this way, this was a system design that clearly delineated Australis' components, functions and interfaces, even if it was not formally documented.

One aspect that still puzzles Owen is how we arrived at the size and mass specification. Perhaps we were thinking that it should be about the same size as OSCAR 3. We was never aware of any specific need to minimise size, mass or even power consumption.

At the same time, we needed a reception station at the University to ensure that we had up-to-date information on the flight. This was readily achieved with the equipment we already had for our weather satellite reception, but we had no transmitter capable of commanding Australis, so this was designated to Bill Naughton in country Victoria. Bill had substantial VHF antennas (and possibly slightly over-the-legal-limit transmitter power in the two-metre band), had participated in some interesting amateur radio experiments, such as bouncing a signal off the moon, and was keen to help.

Designers and builders of the components volunteered, or perhaps, more accurately were volunteered, so that we had a team working toward the goal of building Australis. Peter took on the two transmitters and the HI keyer, David the telemetry system and command decoder, Paul (with help from Geoff) took on

Plate 4.4

Plate 4.4 *The two transmitters and HI keyer.*

the mechanical aspects, and Steve, with help from Owen, the orientation and stabilisation system. The only person who was not a student was Les Jenkins, who built the command receiver. He was a keen amateur who often allowed us into his radio shack to hold conferences with OSCAR using his HF radio. Crammed into his shack in the bottom of his garden and surrounded by glowing valves, dials, knobs and lamps, the distorted voices from the single sideband HF radio were often difficult to understand, but we managed to communicate with OSCAR people successfully.

Richard and Owen took on the task of providing supplies for the team. We understood that the space environment for the electronics was quite different from the earthbound environment. We knew that Australis' internal temperature might easily become very high or very low – we had no way of knowing at the time – so designers set themselves a wide operating temperature range. We knew that radiation was likely to be a problem: the Van Allen radiation belts that surround the earth[8] had been discovered but we had no understanding of the problem. In the end, the best quality components we could get hold of were used with the exception of the space-qualified transistors, which we managed to obtain.

There was virtually no budget for the project, so almost everything had to be donated. Richard and Owen therefore undertook a publicity campaign in the press, radio and television. At every opportunity, and as we shall see there were quite a number, press releases were issued and our faces appeared in front of the public at a surprising frequency. The outcome was that, on telephoning the manager of a company that provided something we needed, we were known and even recognised when we visited. Certainly we cannot recall an occasion, even interstate, when we were sent away without presenting our case. Nor, for that matter, can we remember a refusal, such was the interest in space and the interest that we had created in our satellite project.

By the end of 1966, we had a team busily working on their contribution to the project and, considering that each of us had our studies to attend to, the team worked surprisingly well.

* * *

We had no concept of project management but the role was necessary to ensure that steady progress was made. Quite unreasonably, Richard and I pushed and cajoled those "volunteers" to design, build and test their contributions.

* * *

We will now briefly look at the details of each of these sub-systems and stories behind them. While the orientation and stabilisation sub-system was central to Australis' design, we begin with the electronics and in the next chapter describe it and the structure, batteries, ejection and master power switch systems.

* * *

---

8   Wikipedia, 'Van Allen Radiation Belt' at <https://en.wikipedia.org/wiki/Van_Allen_radiation_belt>. Accessed June 2017.

TRANSMITTERS

As we read in the previous chapter, Peter was a member of the University of Melbourne Amateur Radio Club and experienced in building radio frequency transmitters and receivers. When Owen worked with him later, he seemed to have a happy knack of understanding what were complex RF circuits. The amateur radio club and MUAS met together from time to time to discuss their hobbies and it was during one of these meetings when he met Richard and Owen. Also at one of those meetings the idea of building a satellite was raised and he was interested.

To begin with, Peter undertook to build the two transmitters and the HI keyer. Peter was one of the two PhD students in our group. He had started his research, which entailed studying the ionosphere with others in the Physics RAAF Department using radio methods. His professor was the kindly and dedicated Vic Hopper. Notably, under him was Jean Laby, daughter of an early professor of Natural Philosophy at the University and author of the Kaye and Laby logarithm and trigonometric tables that mathematics students then used every day. Vic Hopper's department, that Owen was to join later, was physically and organisationally separate from the "main" physics department that occupied the Natural Philosophy building where our antennas and meeting room were located. Physics RAAF provided teaching staff for the Royal Australian Air Force Academy at Point Cook.

Peter's early work on his studies required him to build transmitters and receivers to probe the various layers of the ionosphere and he had an affinity with radio frequency electronics. He built the electronics that worked beautifully. Owen worked with another man later in life who likewise had a great affinity with electronics. Whereas Owen had to calculate component values, he somehow simply knew the right value and, invariably, he was correct. Just as well, as he didn't know how to calculate them! Returning to Peter, he was the ideal person to build the two transmitters. He had built similar transmitters before and understood the technology very well.

The High Frequency (HF) transmitter operated in the ten-metre amateur radio band at 29.450 MHz and was amplitude modulated by the audio signal from the Telemetry System. The output power was nominally 250 mW, which was a relatively heavy drain on the battery and so it could be commanded on and off by the Command System. There were a mere three silicon transistors in the transmitter, an oscillator at the transmit frequency, a modulator and a class C power amplifier. However, the key to the transmitter was not so much the transistors but the filtering to ensure that spurious signals were suppressed and the desired 29 MHz signal was delivered to the antenna at the required power level.

Like all the electronics, it had to operate in a wide, but unknown, temperature range. While we were able to specify the battery voltage, the battery would discharge eventually. Accordingly all the electronics should operate to as low a battery voltage as reasonably possible, the transmitters all the while observing the specifications for spurious signals so as not to interfere with receivers on other frequencies. Interference with other electronics within Australis was also a concern. This was no easy task, especially for the transmitters, and you can see the many filter coils and capacitors in the photograph.

As with the other electronics, prototypes were manufactured and tested to satisfy ourselves as much as possible that they would work correctly in flight. But more about testing later. The prototypes were important for another reason. All the flight components had to be accommodated within the satellite structure so the prototypes were useful for understanding how they would be packed into Australis' mechanical structure – no three-dimensional computer modelling then!

Being the RF guru he was, Peter also built the two-metre, Very High Frequency (VHF) transmitter. Though conceptually simpler not having to be commanded, it was to remain turned on throughout the flight: the much higher frequency was a challenge, if only because silicon transistors had reached those frequencies only relatively recently. On the other hand, the power output was nominally 50 milliwatts at 144.050 MHz. Like the HF transmitter, it was amplitude modulated by the Telemetry System.

The design was more complex, however, with four silicon transistors. The 36 MHz crystal oscillator provided a signal at one-quarter of the transmit frequency, which was doubled and then doubled again before being amplified in the final stage. The first frequency doubler was also modulated by the telemetry system. Again, the photograph of the prototype clearly shows the filtering coils and capacitors.

It is worth discussing two concerns that we had. First, the value of the inductance of the coils used in the filters was directly related to their ability to remove unwanted frequencies as the resonance frequency is set by the inductance and capacitance values. Changes in either change the resonant frequency and thus the filtering performance. It was understood that there were two reasons why the inductance and capacitance values would change. One cause was temperature changes, which result in (among other things) dimensional changes in the components. For the inductance coils in particular, the inductance is determined by the physical dimensions of the coil and so, as the metal expands and contracts with temperature changes, the inductance changes. Likewise, the filter capacitors also change values. This was a considerable design challenge for Peter and something that was discussed at our lunchtime meetings in the club's rooftop room.

Plate 4.5

Plate 4.5 *Hi Keyer boards before assembly.*

Peter built the VHF transmitter for AO5 using ferrite core inductors to allow the inductance to be tuned to the correct value whereas the HF transmitter used open coil inductors and high quality adjustable capacitors to tune the circuits. The transmitters were then "potted", that is, filled with epoxy resin, to ensure nothing moved and hence frequencies remained as constant as we could make them.

Both transmitters and the command receiver used quartz crystal resonators, known simply as "crystals", to set the transmit frequency. However, quartz crystals, and hence the transmit and receiver frequencies, do not remain exactly as their advertised frequency because their temperature changes – they have a temperature coefficient of frequency. Different types of crystals have different temperature coefficients and some even have a point where the coefficient is zero but all change frequency slightly with large swings in temperature. It all depends on the way in which they are cut from the original piece of quartz. Care was taken to specify the appropriate crystal type, or cut, for the frequency and temperature range we anticipated for Australis. Here was another task for the "scroungers" of components.

As if he had not enough to do, Peter also built the HI keyer. Unlike his radio frequency work, digital logic was something that Owen could understand – like the stabilisation system, it was a subject that he had just learned about in lectures. The HI keyer was somewhat more complex than the transmitters but

4.6

4.7

Plate 4.6 *Peter's Hi Keyer.*   **Plate 4.7** *The Command Receiver.*

only in the sense that there were some fourteen transistors and about thirty five resistors. The design consisted of a four hertz multivibrator (oscillator) followed by four dividers that reduced the frequency to one-quarter hertz plus four logic gates to produce the Morse Code modulation signal of dot, dot, dot, dot, pause, dot, dot for HI.

The earliest microelectronic components consisting of two or more transistors and other electronic components in a single package, or chip, were beginning to appear around this time and they were Resistor-Transistor-Logic (RTL) logic chips. It was decided not to use them as we did not know whether any were space qualified. In fact, the logic form that Peter chose for the HI keyer was RTL but implemented only with discrete transistors and resistors. Throughout the years more, robust forms of logic have been developed and RTL has followed the dinosaurs into oblivion.

Unlike his transmitters where layout was critical, the layout of the HI keyer was not and a more important criterion was to minimise the volume (and power) that it occupied. Peter chose a "cordwood" construction which was a common technique then, especially in space and military applications. Cordwood layers small circuit boards in three dimensions with wires connecting the various layers. The advantage is efficient use of volume but manufacture is difficult and repair near impossible, unless the fault happens to be on one of the outer layers. Nevertheless, the HI keyer was built up to a block of circuit boards, transistors, resistors, capacitors, wires, nuts and bolts.

It was decided for some reason that the HI keyer should also be "potted". We can only surmise that we thought it would be more robust that way. It fell to Owen to pot it, which he did. He can reveal here for the first time that it included the name of the University college where he resided in from a letter head with the epoxy resin potting. A small part of Trinity College made it into space!

4.8

Plate 4.8 *Dave's Command Decoder boards before assembly. Note the large inductors and capacitors for the ON, OFF and ENABLE resonant audio circuits in the centre upper row.*

* * *

Telemetry Subsystem

The next electronic sub-system was the telemetry encoder, which delivered temperature, etc information to the ground by modulating the two transmitters. The first consideration was what information we wanted. One of the important questions we were seeking an answer to was whether a satellite could be stabilised and oriented in orbit and so we needed to know how Australis was moving once it reached orbit and how that movement, or tumbling, changed with time.

At a lunch meeting where we wondered how we could possibly measure the tumbling. Richard, recalling instruments on various satellites, including the Apollo spacecraft (and especially the moon lander), suggested a star tracker. Now a conventional star tracker is a telescope with a light detector and motors to turn it to focus on the chosen star – the early American interplanetary spacecraft used the star Canopus. This would have been much more complex than the rest of Australis. Nevertheless, his suggestion led to the notion of semiconductor light sensors that we expected to record the transition that the sensor viewed from the blackness of space to the brightness of a sunlit Earth. We later discovered that this was exactly how the weather satellites we were tracking monitored their deliberate rotations.

Three "horizon sensors" were deemed necessary to understand Australis' tumbling, although in truth we had no idea how we would be able to understand the tumbling motion from the horizon sensor signals, but that was a problem for later. An interesting development occurred in flight, however. Not only were the horizon sensors able to detect space to Earth transitions, which they did very easily, but variations in light were seen as the sensors' view passed over the Earth. These were clouds – our own very crude weather satellite, but this was to be some years later.

So, three horizon sensors were to be buried in tubes to restrict their view and shade them from extraneous light. This led to a concern for the potential for damage and destruction should a sensor see sunlight directly. A balance had to be struck between a reasonable field of view for the horizon sensors and a narrow field to reduce the risk of damage from the sun. This was a risk that we had to accept but, again, this was something we had to wait for some years before we learned whether the risk was worth taking. In fact, no damage occurred during the flight.

We now had four channels of telemetry decided; HI was the first telemetry channel and used as an identification and synchronisation channel, for without the clearly identifiable HI greeting, it may have been difficult to identify the other telemetry channels. Then there were to be the three horizon sensors.

We knew that maintaining the temperature of a satellite within a range suitable for the electronics is critical. We have discussed frequency stability of the transmitters and receivers but changes in other electronic components with temperature also affect their quantitative value and hence the way they perform, so temperature of the internal parts of Australis was a major consideration for the design. Accordingly, two temperature sensors (which deliberately have large and deterministic temperature coefficients), internal and skin temperature, were specified. Finally, battery voltage and current measurements were added, making a total of seven telemetry channels plus HI. (Any power of two is relatively easy to achieve in binary logic electronics.) The eight telemetry channels were:

1. HI identification
2. Battery current drain
3. X axis horizon sensor
4. Battery voltage
5. Y axis horizon sensor
6. Internal temperature
7. Z axis horizon sensor
8. Skin temperature.

The Telemetry System connected each of these eight telemetry channels in sequence to an audio oscillator whose frequency varied with the voltage of the sensor connected to it. The output of the audio oscillator modulated the two transmitters. However, being amplitude modulated, the transmitters transmitted more power, and hence drew more current from the battery when the modulation voltage was high, and so the transmitters were modulated by signals with opposite phase so that, while one was drawing maximum power, the other was drawing minimum current, thereby smoothing the overall current drain on the satellite's battery.

Each channel was sampled for 6.5 seconds for a 52-second telemetry cycle. The audio frequencies ranged from 400 Hz to about 3 kHz. For each of the battery voltage, current and temperature telemetry channels, there was a calibration graph provided in the Users' Guide to convert the audio frequency into the parameter being measured. More on the Users' Guide later.

Minimising temperature variations was also an issue for the telemetry system since changes in temperature caused changes in electronic component values that in turn changed the audio frequency. Design choices minimised temperature coefficients of the critical electronic components.

* * *

Plates 4.9 *HI keyer being tested.*

COMMAND SYSTEM

Recall that the ten-metre HF transmitter was designed to radiate five times more power than the two-metre VHF transmitter and hence it would draw more from the battery. It was felt that a system that would turn the HF transmitter on and off would not only allow us to manage the battery current drain, and hence battery lifetime, but also demonstrate for the first time that control of an amateur satellite could be achieved. The two commands required were straightforward – turn the HF transmitter on and turn it off – but its implementation was not simple.

Conceptually the commands would simply consist of two audio tones carried on a two-metre signal to turn the transmitter on and a different pair of tones to turn it off. To simplify, one of the tones, known as the Enable tone, was shared between the two commands, the others being the On and Off tones. The Command Decoder used the three audio tones to determine the state of a bistable flip flop that controlled the transmit state of the HF Transmitter.

There was concern that noise from the command receiver may activate the commands and so it was necessary to ensure that only a narrow range of audio frequencies transmitted for sufficient time would be required to command. The problem of maintaining frequency stability of the three audio filters in space over the expected temperature range was significant for the technology of the day which was narrow band (technically, a high Q factor) resonant circuits with large, and temperature sensitive, inductors and capacitors. Dave managed to achieve these demanding requirements with another cordwood module.

The only designer and builder of Australis who was not a student was Les who built the complex command receiver. Technically, it was a double conversion, superheterodyne receiver with a pre amplifier, two intermediate frequencies, amplitude modulation detector and automatic gain control. If complexity is measured by the number of transistors, it was relatively complex; there were twelve.

The frequencies were deliberately not published in the Users' Guide to ensure that valid commands were only sent by authorised stations but they can be revealed here. The receiver operated at 147.850 MHz in the two-metre amateur band. The ON command required the Enable tone of 2,960 Hz and the ON tone of 3,494 Hz to be transmitted simultaneously for several seconds. The OFF command required the Enable tone and the OFF tone of 3,226 Hz to be transmitted. The bandwidth of each of the command tone audio filters was 30 Hz.

Other than the master power switch, these were the electronic sub-systems on Australis.

Before turning to the orientation and stabilisation system, as well as the mechanical aspects of Australis, it is worthwhile discussing the methods used to make the electronics for Australis. The standard circuit board material used in industry in those days was bakelite, about 1.6mm thick and they were supplied coated with a thin film of copper. The boards were drilled and then, using photographic processes, the layout of the desired electrically conducting tracks was printed onto the board, which was then chemically etched to leave the track pattern. The track layout, which printed a material that was resistant to the etching material, was called "resist". After etching away the unwanted copper, the desired tracks connecting the various electrical components remained.

As poor students, we didn't have the resources to use those expensive production techniques. For the AO5 spacecraft, we used fibreglass for its superior strength and we painted the track pattern on the copper with nail varnish which was our "resist" material. The boards were then etched in ferric chloride to dissolve away the copper where there was no nail varnish. It was not uncommon for people using this technique to return home with holes in their clothes from the acidic ferric chloride. Goodness only knows what the fumes did to people's lungs. There were lots of things to go wrong, especially the etching, and often boards had to be remade and processes repeated.

Plate 5.1 *Partially assembled satellite. The orientation magnet is clearly visible on the left, the batteries are in the centre of the satellite and the isolation diodes at the bottom of the picture. Notice also Paul's meticulous dressing of the wiring loom. The mountings for the upper panel of the satellite are glued to the batteries.*

CHAPTER 5
Australis Mechanicals, Bending the Metal

Perhaps a better title for this chapter would be Australis' Non-Electronic Systems – a bit of a mouthful but more descriptive.

We begin with the Stabilisation and Orientation Systems. We expected Australis may be placed into orbit and ejected from the launch rocket with a spin. Recall that it is desirable for a transponder satellite, and indeed any satellite transmitting radio signals, to have a reasonably steady signal at the receiver so as to avoid "fades" where data is lost or compromised. This is as true today as it was in Australis' time, notwithstanding modern digital communications, perhaps even more so when a weak signal can cause complete loss of digital data whereas a weak analogue signal may only cause loss of measurement accuracy.

Magnetic Attitude Stabilisation System (MASS)

Most large, modern satellites are stabilised and oriented in three directions using a complex set of sensors, reaction wheels and gas jets. This was out of the question for Australis, if for no other reason than we simply could not afford the electrical power necessary to run such a system. No, the stabilisation system had to consume minimal electrical power, or ideally none. Fortunately the Earth's magnetic field, although weak, afforded us a means of crudely stabilising and orienting the satellite.

There are, therefore, two parts to the subsystem – a means of reducing any residual spin of a satellite as it enters orbit and another means of orienting it in a direction suitable for the antennas. The subsystem was called the Magnetic Attitude Stabilisation System, MASS, consisting of orientation and stabilisation components.

Stabilisation Component

Equally fortunate was that Steve and Owen, as electrical engineering students, had recently attended Physics lectures where the physics of magnetic hysteresis had been taught. When magnetic materials are subject to changing magnetic fields, energy is absorbed by the material as the atoms (or rather the magnetic fields of the atoms) within the material are re-oriented by changes in the magnetic field. In some materials, used for electrical transformers for example, this energy is small but in others much larger. Could a material with a sufficiently large "hysteresis loop", as it is known technically, absorb the rotational energy of Australis? The size of the hysteresis loop is a measure of the energy absorbed by the magnetic material for each reversal of the magnetic field, so a large hysteresis loop, absorbing a relatively large amount of energy, is desirable for this task.

After consulting some technical books in the University's library, Steve reckoned that it might just be possible to markedly reduce any satellite spin to a reasonable value in a few weeks. Any spinning or rotating could never be totally eliminated because, as the speed of the spinning was reduced, so the number of magnetic field reversals, and thus the rate of absorbing rotational energy, drops. Nevertheless, reducing the rate to, say, one rotation or fade per minute would make communication far more feasible.

Eddy currents induced in Australis' aluminium skin due to the changing Earth's magnetic field as the satellite rotated would also absorb energy and contribute to reducing the spin. Unfortunately, we were unable to calculate the likely contribution of this energy absorption mechanism.

In addition, we had no idea what the rate of rotation might be. Launch rockets typically rotate about their longitudinal axis to help stabilise the direction that it is pointed. Also, when Australis was ejected from the launch rocket, any imbalance in the ejection springs would add a further rotation, about a different axis. We had to live with the notion that Australis would be spinning when placed in orbit at a rate of perhaps a rotation every few seconds. It scarcely needs saying that this made the design of a stabilisation system and prediction of its performance difficult at best. Furthermore, at the time we had only a vague understanding of the likely mechanical properties of the satellite.

Nevertheless, we sought out a material with as high a magnetic hysteresis loop as we could find and measured the loop. We found an alloy consisting of about 50 per cent iron and 50 per cent nickel with a very high hysteresis loop and made by Allegheny-Ludlum Manufacturing in the US. A local company would import it for us at a cost of US$35, with us additionally paying for the freight from Dunkirk, New York, and customs clearance. This was one of the few components of Australis that we would pay for.

There was one further problem, however. A web page on Alloy 4750[1] states that '… in order to achieve the maximum softness and optimum electrical and magnetic properties, Alloy 4750 should be annealed in a dry hydrogen atmosphere at 2150°F for 2-4 hours, followed by a furnace cool at 100-200°F per hour down to 800°F and at any rate thereafter'. Finding someone willing to heat hydrogen to high temperatures would be a challenge! Nevertheless, that brave someone was found at the Engineering Faculty of Monash University[2] and the wires were duly soaked in a hot hydrogen atmosphere for the due time and allowed to cool at the due rate. The challenges of building a satellite! (We wonder if such facilities are even available today in Australia and if there is anyone willing to undertake the task nowadays.)

Following the annealing, there was a memorable afternoon as Steve and Owen set off to the student's electronics laboratory in the Engineering Faculty at Melbourne University to make the measurement of the hysteresis loop. Arriving at the laboratory with the material in the form of wires about 3mm diameter, an irate and insistent Senior Lecturer Jim P. demanded that Owen not enter wearing thongs

---

1   Aircraft Materials, Nickel Iron Alloy 4650, Technical Data Sheet' at <https://www.aircraftmaterials.com/data/electronic/4750.html>. Accessed June 2017.

2   Incredibly, Fate and rabbits recently introduced Owen and Bruce, the student builder of the Monash furnace.

– imagine the danger when measuring a magnetic hysteresis loop when wearing thongs! (The voltages were low.) Nevertheless, the rules were the rules and Owen trudged off to change into something more to his liking, and in accordance with the rules. Eventually, the hysteresis loop was displayed on an oscilloscope and the numbers recorded and analysed. This is not a difficult measurement to make (Owen has done it a few times since) but it was a novelty then, and a satisfying one, when done for the first time.

The upshot was that yes, with a handful of these wires, the likely spin of the imagined Australis could be reduced to a reasonable value in a reasonable time. With all the unknowns, any quantitative analysis and estimate of the spin-down time was guess work. It would work. Thanks, Jim for letting us use the lab.

After this challenging and very interesting experience, we had our stabilisation system – a bundle of innocuous wires. Better than any lecture or practical class at University, we learned and thoroughly understood the principles of magnetic materials and hysteresis loops.

Significantly and most importantly, the wires required no electrical power for them to do their job; they were passive.

Orientation Component

The ideal orientation system would orient Australis in a fixed direction relative to the Earth, so that the various antennas could be arranged to maximise signals towards the Earth. As we've seen, an active orientation system was utterly impractical for us. We needed another passive system and, once again, the Earth's magnetic field offered the solution.

As any child knows, two magnets attract and align themselves. The Earth's field is one of those magnets, so all we had to do was to place a magnet in Australis and, hey presto, our little satellite would be aligned with the Earth's magnetic field. Since the Earth's field is very roughly parallel with the Earth's surface over much of the populated regions (but not at the poles), here was an easy solution.

Not quite so quickly. How big a magnet is required to follow the earth's

Plate 5.2

Plate 5.2 *The bottom plate of Australis showing Rod's four locating pins, two holes for the Henderson springs and the centre hole for the HF antenna mount.*

Plate 5.3

magnetic field, especially given that the field dips down at the Earth's poles and a satellite tends to do a 180 degree tumble as it passes over each pole? Furthermore, we had introduced a permanent magnetic field which may affect the hysteresis of the stabilisation wires. These were serious considerations that had to be understood and dealt with. More delving into technical books dealing with magnetic fields where much of the theory is contained in Physics and Mathematics that was well beyond our understanding at that stage of our University careers – vector calculus and the like.

Nevertheless, Steve was able to convince himself that magnetic orientation was feasible.

The Rola company in Melbourne manufactured audio loud speakers. Perhaps they would be willing to supply us with magnets, and they did.

As previously mentioned, we had little understanding of how Australis would be spinning as it left the launch rocket.

EJECTION SPRINGS

On advice from OSCAR, we needed to eject Australis from the launch rocket. Further, had we remained attached to the rocket, the opportunities for mounting our three antennas would have been very much reduced, and in any event, the usefulness of a satellite periodically shadowed by the rocket would be very limited. The simplest approach was to use a pair of springs to push Australis away from the rocket. This required the rocket to have a mechanism to hold Australis in place before and during the launch, and then release it – but that would be the responsibility of the rocket people. We simply had to provide the springs and a locating mechanism.

The springs had to be matched, that is, each of the springs had to have the same force characteristics as they were compressed and released so that Australis would be pushed evenly into space with a minimum of spin imparted by the ejection. Persuading the Henderson Federal Spring factory to supply them, which it did, is a story related in the next chapter.

Plate 5.3 *A prototype of the antenna mount made by Rod. The antenna really was a tape rule!*

Antennas

Three antennas were needed, one each for the two transmitters and one for the command receiver. The two, two-metre VHF antennas were to be tuned monopoles with the structure of Australis acting as some sort of ground plane, whereas the single ten-metre HF antenna was a tuned dipole with the satellite structure itself in the middle of the antenna. The two VHF antennas were estimated to be about 475 mm (19 inches) long and the HF antenna roughly ten times as long, with each arm being about 1,300 mm. It was unlikely that any rocket would be able (or the organisation willing) to accommodate a satellite that, with antennas, measured at least 2.5x1 m, so the antennas had to be folded into a more compact package.

A simple solution was suggested – flexible carpenter's rulers that were simply folded around the package and that would unfold under their own once released. Simple. And again, passive. We were concerned that the less-than-ideal conductivity of the steel might cause signal losses but experimentation showed that it was insignificant. Turner Industries in Melbourne made hand tools for wood and metal working, and steel tapes. Owen still has a piece of steel tape that was cut off from the end of one of the antennas during the tuning process. And he still wonders at the piece at the other side of the cut that is still wandering daily around the Earth with its feet, inches and centimetres marked on it.

Tuning the antennas was another sign of the times. Almost the entire crew trooped down to Les' home in Clayton, Melbourne where he had the equipment to tune the antennas. It was a beautifully sunny day with Les' wife sunning herself on a sun lounge. Concentrate on the job, boys!

Les had a radio signal strength meter which we used to maximise the signal but, of course, it was changed by the presence of anyone nearby reading the meter. Never mind, Les brought out his 22 gauge rifle with its telescopic sight so that we could read the meter some distance from it. As if to prove this, Owen have a picture of him looking through the sight at the meter.[3]

Battery

The Australis batteries were an alkaline manganese chemistry chosen for their slow self-discharge rate and relatively high capacity. They were a new type of battery then, with the patent only having been granted in 1960. The chemistry is the same as the modern "alkaline" battery available in supermarkets and stores everywhere. However, each cell in Australis was considerably larger and enclosed in a cylindrical steel case.

The terminal voltage of alkaline cells is 1.5 volts and so 14 cells wired in series made a single 21 volt string. There were two such strings, each supplying one of the two transmitters. The remainder of

---

3   For disbelieving American readers, guns are very tightly regulated in Australia following the massacre of over 35 people and 23 wounded in Port Arthur, Tasmania in 1996. It would be unthinkable in Australia for a rifle to be used this way today. There will be no discussion about the pros and cons of gun control in this book!

the electronics – that is, the command receiver and decoder, telemetry system and HI Keyer – were supplied by both strings via isolating diodes. In this way, we were assured that one transmitter, telemetry and command system would continue to operate even if one battery string failed. Should one string fail due to an excessively high current drain in one of the electronics systems, for example, the other string would continue to supply its electronics, thereby giving Australis some redundancy. Individual cells were encased in epoxy resin.

Union Carbide supplied the batteries, but due to the long time between construction and launch, the batteries were replaced in the US prior to launch.

POWER SWITCH

Obviously, Australis had to be dormant and not drain its batteries until it reached orbit and was ejected from the launch rocket. Two micro switches mounted on the side of the structure closest to the launch rocket applied power when Australis was released from captivity.

STRUCTURE

In many ways, the structure was the most important single item of Australis as every sub-system required its support. Paul's superb manufacture of the structure amazed us with its perfection – Owen's very amateurish attempts at woodwork were well and truly put to shame by Paul's craftsmanship and accuracy. The electronics captured attention but the less glamorous structure (and stabilisation and orientation sub-systems) were equally important and significant.

The shape of Australis was determined by OSCAR 3 launched in March 1965. We asked the OSCAR people

Plate 5.4

Plate 5.4 *View from above of the almost completely assembled Australis. The VHF antennas are installed (left and right ends), the enclosures for the two springs are clearly visible and the teflon thermal isolation between the internals and the outer skin is visible on the panel at the top of the picture.*

whether that was the right shape and dimensions for our satellite and they replied, "close enough".

The structure was built around the heaviest and largest item: the battery. Electronics boxes were mounted on either side – HI keyer, HF and VHF transmitters on one side and Telemetry and Command Receiver and Decoder on the other. Magnets and hysteresis rods lay on the long side but as far apart as possible – the magnet was attached to the baseplate while the rods were glued to the top sheet of the structure. Antennas were mounted on Teflon pads on each of four sides, horizon sensors viewed space through small holes in each of the defined primary axes of Australis, that is x, y and z axes.

A layer of Teflon was used to thermally isolate the battery and electronics from the outer case which became a thermal shield. The launch authority designed and applied a thermal control coating to the outer case, in this case a partial copper coating and a highly polished metal surface.

Other than the Teflon thermal and electrical isolation, the entire structure was made from an aluminium alloy (1.0% magnesium, 0.6% silicon, 0.2% copper and 0.2% chromium).

Rod wrote:

My contributions to the satellite were the antennas and the four mounting bolts or locator pins. The satellite was located on the launch vehicle using four conical locator pins on the side of the satellite that mated with four matching indentations on the launch vehicle. The specifications for these pins were quite tight and the angle of the cone was critically important. This angle was required to be within one degree of the specification. I made these locator pins by turning an octagonal aluminium rod using a small lathe in my home workshop. The hardest part of the process was getting the angle of the cone exactly right. I did this using an accurate template for the angle and a low-power microscope to make accurate observations of any deviations between the template and the pins.

Following the lead of earlier Oscar satellites, we chose to use metal measuring tape for the antennas. For a first prototype, I bought a number of retractable Stanley tapes at the local hardware store, Carlton Hardware, on Lygon Street, Carlton and removed the tape from the spring-loaded cases. I cut the tapes to length and epoxied them into slots in pieces of Teflon approximately 5 mm thick. These Teflon insulators were attached to the outside of the satellite. For the final version of the antennas, we used tapes provided by Turner Industries.

Plate 6.1 *Preparation for a balloon flight using a weather balloon being filled with hydrogen by Richard on the roof of the Natural Philosophy building. The single helix antenna is in the background.*

CHAPTER 6

Publicity—How We Told Our Story

Somehow, even at boarding school, Owen was aware of the value of well-placed and timed publicity. In those days, large schools had army cadets corps and certainly his school had a large and active corps. Dressed in khaki uniforms each Tuesday afternoon, World War I Lee-Enfield 303 rifles were handed out and the corps paraded around with them before shooting at the practice range. Owen was always amazed at the longest range which must have been several hundreds of metres long, at least; but, disappointingly, he never managed to shoot one.

Owen continues:

> I was a bandsman, improbably tooting on an improbably large trombone. I did enjoy band practice and contributing to the cacophony of our military band. I did not enjoy marching with the band, but the event that was most difficult for me was the annual cadet camp at Puckapunyal army base, north of Melbourne. Being an asthmatic (and at times it was very severe) the idea of sleeping on a "palliasse" – a large hessian bag filled with straw designed perfectly and specifically to be uncomfortable and to stir up asthma – did not appeal to me.
>
> As senior boys were able to elect to do community service in place of army corps, I elected to do so in my final year at school. However, the band leader worked on me and finally persuaded me to continue with the band. I agreed, on condition that I did not have to go to camp. When the time came for the camp, I discovered that the master in charge of the corps had not heard of the arrangement and I had to visit Puckapunyal for the last time. I was not a happy bandsman.
>
> The moment for revenge came when I read an article by a certain Lieutenant Colonel Flood, who was controller of army catering. He had defended army food and, as it happened, I agreed that the food we were provided with at camp was good. I wrote, with impeccable timing, to the afternoon newspaper complimenting the army cooks and staff. I wrote to emphasise the matter, after identifying the school: '… The school has a large and modern kitchen with staff proportionally three times as large as the army staff at Puckapunyal. Even so, the army food is far better than school food.' and I went on to congratulate army cooks.
>
> I had timed the letter to be printed during the final school holiday and two weeks before returning for my final term. To emphasise the matter, the afternoon Herald newspaper ran a two column article, with pictures, titled "Army chow 'up to hotel standard'" and with a photo of Sergeant D. B. Mace in his cadet uniform. (They got the initials wrong – Owen Bullivant Mace, the middle name being my mother's maiden name.)

Surprisingly, no letter arrived at home suggesting that I should not attend the school, nor was anything said on my return, although I did make sure that I was seen as little as possible by masters. Except for the French teacher, who rightly ejected me from his class during that term. Nevertheless, and just to spite him, I managed a pass in the French exam - just. I didn't know then that he was a heroic and exceptionally brave commando in the islands north of Australia during the war and, if I had, he would have received a good deal more respect from me.

I learned three things from this episode: the press can have a huge influence, timing is everything and Jim Glover was a very modest war hero, as were others on staff.

During my first year as a member of MUAS, we had begun collecting weather satellite pictures for the Bureau of Meteorology, as has been described. There were occasional press releases and certainly the Bureau proudly used these new pictures from space in their forecasting, but with not a great deal of acknowledgement of our efforts.

Once we had taken the decision to build a satellite, we realised that we had to become known in the community and so we began a publicity campaign. We were not total neophytes to the publicity game; after all, we had issued press releases through the student's union on a few occasions after listening to various satellites. It seemed that simply issuing a press release on official University letterhead – there was no Public Relations Office then – would result in the press, radio, television and magazine reporters calling us. It should be recalled that space had captured the public's attention. President Kennedy's challenge to put a man on the moon by the end of the 1960s meant that preparations were well underway and various manned space flights were leading up to the flight of Apollo 11 and the moon landing on 12 July 1969. Interest in space was building with these reports, including photos and vision being broadcast onto television sets with almost every news programme. The press were piqued by the thought that even in l'il ole Melbourne there were people dabbling in space, and University students at that.

Our first Australis press release in April 1966 announced our intention of building a satellite and the newspapers responded better than we could have hoped. Owen has articles published by The Sun and The Australian. The front page of The Australian on 15th April carried a picture of four of us clustered around a mock up of what we envisaged Australis would look like – a striped box about 300 mm square with sticks emerging from four sides. How young we looked and how optimistic we were in expecting Australis to be launched by the end of that year. Nevertheless,

Plate 6.2

Plate 6.2 *Launch and away! The parachute and payload are hanging below the balloon.*

there we were, ready to conquer space for Australia.

We were not sufficiently sophisticated in the art of public relations to have a single person designated to speak on behalf of us all, so many did and the press sometimes wanted to interview a particular person, such as Paul, who had a couple of large articles that described his interest in amateur radio. Paul expressed his belief that the requirement for proficiency in Morse Code was outdated – it has since been removed from the amateur license test. He had a World War 2 aircraft radar that he could use to detect clouds at a distance of up to 160 km, anticipating today's weather radars. He made an eloquent argument for communications satellites and video telephones, once more anticipating what we consider as normal today. The Sun recorded that Paul's ambition was to built automatic landing devices for planes, among other ideas.

The Sun carried a brief article noting that we had 'a devil of a time thinking of the right name'. Apparently we rejected 'Boomerang' in favour of Australis on the grounds that it 'would be embarrassing if it decided to come back'. The article reminded us that a large number of names for the satellite were canvassed. Australis seemed to be a natural name for an Australian satellite.

Plate 6.3

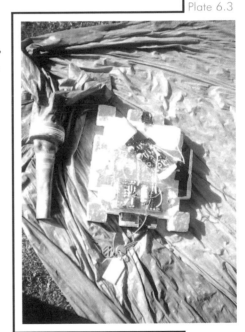

There followed discussions about the strict accuracy of reports and Owen argued that it was not as important as getting our names and faces into the press. His position was that when people read a newspaper or magazine article, hear a radio broadcast or see a television news programme, it is only the broad facts that are remembered, not the technical detail. We should, therefore, aim for our activities and plans to be published frequently and in all mediums.

Although Owen kept copies of all the newspaper and magazine articles that were found, there are none concerning our collection of weather pictures. He has a dozen or more slides of those pictures, however.

* * *

By mid year 1966 we had a clear idea of what we wanted to achieve, and the name we had chosen, and so now was the time for a publicity blitz. Electronics magazine described Australis as being about 430x300x150 mm and weighing around 20 kg. The August edition of the same magazine named the OSCAR team in California that included two we would subsequently meet, Lance Ginner[1] and Harley Gabrielson. It also described the progress of arrangements for the launch and reception of Australis' signals. The daily

---

1   See http://www.arrl.org/files/file/Technology/Bilsing.pdf for pictures of the young Lance.

Plate 6.3 *Our simple payload landed a hundred kilometres east of Melbourne and was recovered. The remains of the balloon with its 'float valve' and my VW keys are visible.*

newspapers also carried articles, some multiple columns with pictures.

* * *

Soon after our go-ahead came from Project OSCAR in 1965, it was decided that we should conduct a number of tests of the various sub-systems, especially the two transmitters. There were at least three reasons, including to publicise the project and to discipline ourselves to meeting a deadline. The third aspect was to exercise ourselves in the logistical requirements to achieve an end, which today is called testing our project management abilities. All these three goals were seen as important and, in reality, they were a mature, professional judgement of what is required for success.

As the reader can imagine, operating a satellite sub-system in an environment similar to that expected in space is not straightforward. We needed a hard vacuum and a solar radiation (heat and light) environment similar to that in orbit. The Army Test Establishment at Maribyrnong in western Melbourne had thermal chambers for testing army equipment, which is used at ground level. It also had vacuum chambers for testing Royal Australian Air Force (RAAF) equipment. Would we even be allowed to use those facilities? None of them could simulate the space environment that we were expecting. We needed to get into, or at least close, to space, but then the question is, how to do so?

Richard's relationship with the Bureau came into play. He had been employed to fly weather balloons for the Bureau. These balloons carried small packages that measured wind, air temperature, pressure and so on and were carried on small rubber balloons to altitudes approaching 40 km. This would be adequate for our purposes and had the considerable advantage of being far more newsworthy than a vacuum chamber in an army establishment. Furthermore, Peter's professor, Vic Hopper, had been experimenting with a means of "floating" weather balloons at a certain, fixed altitude. He achieved this with a simple and ingenious contraption consisting of an aluminium tube inserted into the filling tube at the bottom of the balloon. A spring held a ping-pong (table tennis) ball in place at the inner end of the tube. A string was attached to the ball and the top of the balloon so that when the balloon reached a certain size, the string lifted the ball off the tube allowing gas to escape and thereby stabilising the altitude of the balloon. The key to the contraption working as hoped was the tension in the spring, the length of the string and the attachment of the string to the top of the balloon. All too often these were not quite right and the balloon burst, as they would anyway without Prof's tube. Professor Hopper had flown a number of balloons finding the correct combination of tension, length and attachment, and he was able to float balloons, but not with 100 per cent success, as the balloons were never designed to have a string attached to the apex where, it is true, the latex was thicker and, presumably, stronger. The balloon material became brittle in the cold temperatures and frequently broke.

As an aside, Vic was an experimentalist interested in getting into and above the atmosphere. His research interests were gamma ray astronomy which was conducted on very, very large balloons. He flew cameras on tethered balloons over archaeological sites in Syria, thereby anticipating the use of drones for exactly the same purpose decades later. With his balloon and camera, he helped the Australian archæological team discover features that were not visible from the ground. He proposed large tethered balloons with electricity generators hanging beneath them, anticipating a renewable energy source. Owen's favourite exploit of his was a tethered balloon flown from the University. The tether line broke, but, much to his surprise, the line remained aloft for over thirty minutes before being draped for many hundreds of metres across Melbourne roads, parks and homes. Imagine what the startled motorists and pedestrians would have thought on seeing a string gently settling across their path! Owen's infant son was a favourite of his when Owen joined his department a few years later.

Nevertheless, weather balloons were our chosen means of testing Australis sub-systems and a number of flights were conducted from the roof of the Natural Philosophy building within the University of Melbourne campus, which is barely a kilometre from Melbourne's central business district. There are numerous slides taken on glorious sunny days while preparing for a flight. And one of Owen showing how strong he was carrying a 50kg steel cylinder of hydrogen gas from ground level up two flights of stairs to the roof. 'Trying to calm my heavy breathing and aching shoulder once on the roof, I decided that I would not be so stupid next time and ask for help.'

Plate 6.4 *A test translator package that was flown in February 1968. The radio frequency electronics is sitting on the polystyrene box and a 'baro-switch' to turn on the package above a pre-set altitude sits on the grass in front of the package. Richard is holding the 432 MHz transmit antenna. Amateurs' conversations, called contacts, were successfully conducted through the package.*
Plate 6.5 *A polystyrene package with a HF transmitter and HI keyer was flown in December 1967 on a HIBAL flight.*

There were discussions held on what to call these test flights and we settled upon 'Bravo' for balloon and on the basis that the 'B' should follow the 'A' for Australis, presumably. In any event, three Bravo balloon flights were planned, beginning with the VHF transmitter and HI keyer. The flight was publicised by the WIA, whose Australian headquarters were located in Melbourne, convenient for us. Its signals were heard at Birchip in the Wimmera some hundreds of kilometres from Melbourne.

Perhaps the launch of one of these balloons should be described. Once the day had been decided; it was always a weekend so that amateurs could listen and we students had no academic commitments. Peter arrives with his transmitter and test equipment, checks it out and gives the thumbs up, OK for flight. While the electronics are packed into the polystyrene box, the hydrogen gas cylinder is prepared with regulator and hose attached. Richard, experienced flying weather balloons for the Bureau, begins inflation. Once it reaches about 2.5 metres high and 1.5 m in diameter, the balloon is ready. Remove the filling hose and seal the balloon if it's one without a float device. Attach the radar reflector and, below it, parachute and payload, which is a mere kilogram or so.

Now remember that we are planning to launch this balloon from the University of Melbourne which is close to the city of Melbourne and perhaps ten kilometres from Melbourne's then international airport, the Essendon Airport. Fortunately, we could see aircraft approaching the airport and so it occurred to us that some might not see quite the same excitement in our balloon flight as we did. Indeed, it might even be seen as a "hazard to aviation". Recognising this possibility, we waited for the aircraft to pass before launching. We telephoned the control tower at the airport, as it was then, asking when might be an appropriate time to launch. One of the newspaper photographs even shows Richard with telephone in hand, no doubt deep in conversation with the tower. As unlikely as such co-operation seems today, the Okay was given and we duly launched at the suggested time.

By the time that we were ready to launch, the balloon, radar reflector, parachute and payload stood perhaps as much as ten metres above our heads. Needless to say, low winds were necessary for a successful launch.

The launch consisted of a theatrical count down and then whoever was holding the payload simply let go of it. Richard followed the ascent using a Bureau theodolite until it disappeared in cloud or behind a building. Everyone rushed to our garret in the rooftops to listen to the HI signal and measure whatever we measured on the flight, mostly internal temperature.

If it was a clear day and everything went well, Richard might have been able to track the balloon for a considerable distance. Of course, the balloon expanded as it rose and the atmospheric pressure dropped, making his task easier. In truth, we had no way of accurately, or even inaccurately, knowing where the balloons took themselves, instead relying on farmers to find their remains and telephone the number written on the side of the payload. At least one flight was recovered this way from Avenel, north of Melbourne in Victoria, and another probably landed east of the city around Lilydale.

We were keen to recover our payloads, especially the electronics as there were valuable transistors to be reused. Cars, including Owen's trusty VW and on one occasion his parent's Holden FB station wagon (complete with the fashionable fins of the day), set out with receivers and aerials to pinpoint the payload. We don't believe that we found any of the payloads this way, but it was fun racing around the Victorian countryside, stopping from time to time to listen to the signals while the balloon was still in the air. We knew the general direction of the balloon by using a directional aerial but "pinpointing" would be somewhat an overstatement. A couple of cars could triangulate the position in theory but the beam width of the antennas limited the positioning accuracy. Nevertheless, it was clear the direction that the balloon took and when it burst and the payload was descending. Once on the ground with the aerial lying on the ground, signals were so weak that they were unreadable unless the receiver was very close to it.

Plate 6.6

Three flights were conducted: Bravo 1 carried a test two-metre VHF transmitter with HI keyer and some telemetry. Bravo 2 was similar but this time a ten-metre HF transmitter was flown. The third flight tested the command receiver but by this time the Civil Aviation Authority was becoming nervous about the risk of causing mayhem to aviation – the mayhem was to come some years later.

We were asked not to launch Bravo 3 from the University, so we moved to the Dandenongs, a range of hills – Melbournians call them mountains – to the east of Melbourne. We set off early one winter's morning to the gloriously named Mount Tulle-be-Wong to launch a flight to test the command system.

Despite everything, each flight was successful in that we managed to get all the equipment and people together and, furthermore, our electronics worked satisfactorily. Excellent. However, Richard remembers one flight from the Dandenongs where the VHF signal started to fade as soon as we launched and had disappeared completely a few hundred metres up – much to the embarrassment of the builder!

Importantly also, the press followed us and gave us the publicity we needed and Owen began to record our activities with half half-frame slide camera. We could now press ahead with building the flight hardware.

We will delay the visit to the Potter Foundation until the appropriate moment in the story and follow the construction and testing of the satellite itself, the flight model.

Plate 6.6 *This is a high altitude HIBAL payload, similar to one on which we piggy backed, hanging from the 'hook' on the launch truck with the parachute in front of the truck. The balloon and parachute are tensioned ready for launch. Our package was taped to the main payload on the lower right side. The helium tank for cleaning the filter pads and an air intake funnel rising out of the payload are visible.*

Owen's Photographic Record

Hard to believe these days with our digital cameras that allow us to snap away without concern for the number of pictures taken and that pictures, and movies, were then recorded on film. There were two popular film formats. One approximately 7x10 cm negative film from which prints were made at a size of choice. The other was a "35 mm slide" where the image was recorded in a 35x23 mm positive film. Both required the film, sealed in a lightproof canister, to be taken to a chemist (pharmacy) to be "developed" and prints made, if required.

Following development, 35 mm film images were mounted in 50x50 mm cardboard "slides" ready for showing with a slide projector that, between blown globes, threw the image onto the wall of a darkened room. The advantage of the more modern slide technology was that it was less costly and more likely to be in colour, but viewing required a projector.

With an eye to economy, I had a "half-frame" slide camera that used half a standard 35 mm slide but was mounted in the same cardboard "slide" and gave me twice the number of images per roll of film.

I often chose speed "Ektachome" film, which has not fared so well over the decades with the reds fading somewhat, giving a distinct blue tinge to the images. Resolution of half-frame is half that of full frame and there is evidence that some of the dyes of my Ektachrome images have flowed.

Nevertheless, with Paul's far more professional photographs, these are the majority of the images of those days. I have thousands of half and full frame slides from my late school days to the late 1990s when crude digital cameras began to make their appearance.

There were to be more flights, which, although they took place after Australis had been delivered to the US for launch, were another great adventure. Let's look at one now before moving on with the flight hardware and qualification testing. This is a little known story of nuclear espionage! There is very little recorded in the world wide web.

The independent minded French Gaullist government developed and tested a nuclear weapon in 1960. While early tests were conducted in the Sahara Desert in Algeria, the political instability of the former colony resulted in a relocation to French Polynesia in the Pacific in 1966, and many of the tests were carried out in the atmosphere.

The US wanted to understand the progress being made by the French and it chose to sample the clouds of very fine and radioactive debris, that is dust, from these tests as they spread around the globe. The composition of the dust debris would inform the US of the technical details of the bombs that the French were testing and their destructive size.

The dust rose to extremely high altitudes in the atmosphere, typically from 80,000 to 140,000 feet (24 to 42km). There was only one practical method of obtaining sufficient dust for analysis and that was by flying sampling equipment on very large, high altitude balloons. The balloon technology was well established by a number of companies, including Raven Industries in Sioux Falls, South Dakota (which Owen visited some years later), Winzen Research, Minneapolis, Minnesota and others. Using the cover of the National Science Balloon Facility (NSBF in Palestine Texas) and the US Air Force, the Central Intelligence Agency (CIA) regularly flew a series of flights at altitudes ranging from about 80,000 feet to 135,000 feet. Flights were conducted in South America, Southern Africa (we believe) and Australia. By the time the cloud reached Australia, it had travelled eastwards almost completely around the world and, presumably, dissipated somewhat, but even the dissipation might reveal something of the tests.

The large payload consisted of a pump that forced large volumes of high altitude ambient air through a series of fine filters which collected the radioactive dust. The filter pads had to be recovered after the flight and returned to the US for analysis. Flaps and valves of the dust collection system were operated by a command system to ensure that the filter pads were not contaminated by dust on the ground and at lower than planned altitudes. There was what we thought of as an extensive telemetry system to measure air flow, altitude and other parameters of the flight. Position was measured surprisingly inaccurately by today's GPS standards using the Loran C navigation system.[2] Once at altitude, valves and flaps were opened and the air pumps turned on for the few hours required to capture sufficient dust. Once the pumping was complete, valves closed, pumps turned off and the balloon "cut down" using the command system. A low frequency signal, below AM frequencies, was then turned on to track the payload down to Earth and allow recovery of the payload. Being such a low frequency, a very long antenna, some hundred metres, was released at balloon cut-down. The low frequency ensured that the

---

2   Wikipedia, 'Loran-C' at <https://en.wikipedia.org/wiki/Loran-C>. Accessed June 2017.

signal could be heard even with the antenna lying on the ground.

The launches of Project HIBAL, as the flights were called, were carried out from the balloon facility at Mildura, on the Victorian border with NSW where a Piper Aztec twin engined aircraft was based. (Detail: the aircraft's registration was VHPOU.) The aircraft followed the balloon and its descent to the ground and then directed recovery crews to the payload. As a lead in to future chapters, scientists from around the world had realised that this was a relatively cheap means of lifting their instruments to near space.

The MUAS team also realised that this was an opportunity for flights longer than we could expect from a weather balloon, and probably higher altitude also. In December 1967, six months after we had delivered Australis to the US, we flew an amateur band transponder on a HIBAL headquarters of the Australian Balloon Launch Service (ABLS) at Mildura airport. On this occasion Owen took 35mm slide photographs, fortunately not with Ektachrome, recording the preparation, launch and recovered package. The photographs clearly show our package as well as the dust sampling equipment. Even his VW features.

There were four further HIBAL flights in the first quarter of 1968, successfully testing translator electronics[3] and there is a fax of an image transmitted through the translator.

* * *

Let us now relive a typical such balloon flight, such as we flew the test in December 1967, as they were to be part of Owen's professional life for many years when he was engaged in balloon-borne infrared and gamma ray astronomy, noting that the atmosphere absorbs the infrared and gamma ray radiation that he wanted to measure. These are unmanned flights using very, very large balloons, some as much as 100 metres in diameter at altitude.

Owen remembers:

> After a year, perhaps many more, the experiment is ready to be flown. That is, it has passed all the tests in our laboratory and has been calibrated, proving that it measures whatever it's supposed to and that it survives the pressure and temperature environment at altitude. The balloon has been delivered to the launch site and the launch crew at Mildura, Victoria or NSBF, Texas, USA who have been contracted to fly our experiment.
>
> The experiment and test equipment are packaged and shipped to the launch site, which is almost always a regional airfield – the places I've flown from include the NSBF headquarters at

---

3   Receive on the two-metre VHF amateur band and transmit on the 432 MHz Ultra High Frequency (UHF) amateur band.

Palestine (pronounced "palest – een"), Texas and Malden, Missouri in the USA. In Australia, the towns were Mildura, Broken Hill, Longreach and Alice Springs.

The time of year is carefully chosen to give as long a flight as possible to maximise observation time. The stratospheric winds at the altitudes that the balloons reach change direction twice a year, in March and September approximately, so flights are made when, hopefully, there is little or no wind at altitude.

Once again the experiment is tested and then integrated with the command and telemetry package typically (but not always) provided by the launch crew. Once everything is ready, and having regard to the forecast weather conditions, a launch day is selected.

But first, we must understand something about the balloon itself. It is incredibly large and fragile. It is made from a specially formulated and thin plastic sheet, similar to the plastic that covers dry cleaned garments after they have been cleaned. Strips of special plastic up to 100 metres long and two metres wide, called gores, are shaped and then heat welded with glass tape (for strength) in the seams. Top and bottom plates are attached to the tapes and, crucially, a rope passes through the bottom plate to one gore; its purpose will become clear shortly. Once complete, the balloon, now a very long bundle of gores, is folded carefully into a large box for shipping to the launch site. The boxes can be quite large, depending on the size of the balloon, up to six metres long, two metres wide, two metres high and weighing up to a tonne or even more.

Once inflated, the balloons are at the mercy of the slightest zephyr of wind and so dawn launches are much preferred as that is the time when the wind is at its lowest. Well before dawn, the equipment and balloon are taken out onto the airfield, and a long canvas sheet laid out downwind and carefully swept as the balloon will be laid out on this sheet. Pebbles, bindies and the like can easily puncture a balloon.

The balloon is now laid out along this sheet with its top end under a special roller. Once the experiment has been covered with insulation and the launch team's equipment attached, it is hung from the "pin", high on a mount at the back of the truck. The launch truck backs up to the bottom end of the balloon whereupon it is connected to the payload by the parachute. The long rope that is connected to a gore at the top of the balloon is tied to the parachute.

At the top end of the balloon, which can be as much as 100 metres away, helium gas is pumped into the "bubble" which soon stands up high above the roller holding it down. More and more gas is poured into the bubble from huge ten metre long cylinders on a semi-trailer until the carefully calculated and checked volume has been delivered.

To paint the scene, recall that the launch is timed to occur shortly after dawn, so the events so far have taken place in the dark with lights and torches darting, men in white overalls nervously marching around swiftly and the "scientists" looking on, hoping for a good flight. As dawn breaks, the purples, reds and orange colours appear in the sky. Seen through the plastic of the "bubble", dawn at a balloon launch is a memorable sight indeed.

Owen continues:

> Surprisingly, what is even more memorable is the noise – let me paint the picture, if I can, of the launch, but first understand that the payload is held onto the launch truck by ropes that will be cut at the appropriate moment by explosive cutters, a .22 cartridge that drives the cutter.
>
> As the launch approaches, more and more shouted, even walkie-talkied, commands are given: 'Check this, check that, is everyone ready? Yes.' Trucks that have been idling for hours are revved amid clouds of diesel smoke to ensure they are ready to accelerate rapidly. Finally, the launch director gives the word to launch.
>
> Bang. No, a very loud metallic crash as the roller is released and the lift of the now-unleashed balloon rams the roller into a catcher.
>
> The rustle of the balloon plastic overtakes other noises as the bubble lifts the length of the balloon off the canvas and into the air.
>
> Then the launch truck roars into life and takes off after the "bubble", aiming to get underneath it before it passes over the airport boundary. The aim is to get the payload below the "bubble" so that the payload is lifted off the truck and does not swing into the ground. But look, the truck is veering off to the right – the wind has changed a little!
>
> As the truck disappears behind its dust cloud with the balloon standing majestically above it and the noises recede, suddenly there are two loud bangs and the payload rises.
>
> A successful launch.
>
> And now the work really begins monitoring the experiment, sending commands and getting excited by the results flowing in, but not before we've watched the balloon rise and recede into the distance. Using binoculars, we can see that the helium gas is expanding

Plate 6.7

Plate 6.7 *Filling the 'bubble'. This is a relatively small (but still huge) balloon as there is only one filler tube. Note how the balloon is held under a roller on the trailer, and also note the distance to the parachute, launch truck and payload.*

and filling out the balloon. In time it will reach float altitude once the gas has filled the balloon and excess flowed out its open bottom fitting. If we are lucky enough, we may be able to see it at float, looking more like a tiny upside down onion than anything else, with the sun glinting off the plastic.

After some hours, as the balloon heads towards the horizon, the Piper Aztec plane VHPOU takes off to continue tracking. For whatever reason – batteries dying, getting close to the coast, whatever – eventually it is time for the flight to be ended and so the cut-down command is sent. While we cannot hear the explosive rope cutters that cut the rope between balloon and parachute, it separates them and the cord reaching to the top of the balloon rips out a gore, and the gas escapes, turning the balloon inside out. Remarkably perhaps, the balloon, now more like a candle, reaches the ground well before the payload on its parachute. Both are spotted by the plane which guides the recovery crew.

That's the theory and mostly that's what happens but, as the reader might imagine, there can be incidents and disasters.

It was just this type of flight that we flew our December 1967 test from Mildura airport.

* * *

1966 and 1967 was when our efforts to publicise the Australis project began to pay off. We began to target suppliers of the components we required. Most were very generous indeed and supplied us with whatever we asked for and they were acknowledged in the Users' Guide described later. The list is, in alphabetical order:

- Acme Engineering, Melbourne gave us radio frequency connectors.
- Cannon Electric Ltd, Melbourne provided resistors and connectors. The resistors (electronic components) were high stability, high accuracy tin oxide components not then normally used in ground based electronics.
- Ducon Condensers Pty Ltd, Sydney supplied all the capacitors (also electronic components) used in Australis. Certain capacitors defined the audio frequencies we used in the command sub-system and so their temperature sensitivity was critical. They passed the test.
- Fairchild Australia Pty Ltd, Melbourne provided all the transistors that were used, including very expensive space qualified transistors especially imported for us from the USA.
- Melbourne University (Student's) Union gave us a generous grant for ground equipment. This was probably the $400 of cash that Owen recalls.
- Plessey Components Group in Sydney donated a travel grant, although I have no idea how this came about.
- The Potter Foundation in Melbourne gave us two travel grants. More on this wonderful incident later.

- Pye Pty Ltd, Melbourne provided all the radio frequency crystals that set our two transmitted and command receiver frequencies.
- Rola Co. (Aust) Pty Ltd in Melbourne supplied the magnets for the stabilisation sub-system
- Sample Electronics, Melbourne supplied circuit boards.
- Turner Industries Ltd, also in Melbourne, provided the satellite antennas.
- Union Carbide Australia Ltd., Melbourne, supplied flight and backup cells for the battery packs.
- The Wireless Institute of Australia gave us a generous grant for running expenses.

There are some interesting observations to be made from this list. First, the majority of these suppliers were in Melbourne because it was the centre for manufacturing in Australia then. It could be said that Sydney had a strong electronics manufacturing capability with STC, for example, and that is correct, but it was only natural for us to approach Melbourne offices of companies wherever possible. While some represented US and UK companies, there is a strong representation of Australian companies. It's probably fair to say that few, if any, exist today and that Australia has given up much of its manufacturing expertise to overseas companies. But enough lamentation of this well-known fact – for now.

Plate 6.8

Plate 6.8 *Ready to launch! The launch crew are in white overalls and dignitaries in the foreground.*

We also acknowledged in the Users' Guide the help we received from the Meteorology Department of the University and the Bureau of Meteorology which was, and remains, headquartered in Melbourne. The Glaciology Department of the University also unknowingly contributed low temperature testing chambers: that is, a chest freezer capable of reaching temperatures below domestic fridges. Perhaps it was wise not to acknowledge their unintentional contribution!

Almost without exception, companies that we approached for their contribution did so willingly and, importantly, helpfully. Clearly, our publicity had hit our intended targets, the decision makers in companies.

* * *

There were a couple of notable visits to

AUSTRALIS OSCAR 5

companies to acquire the components we needed and the trip to Henderson Spring Works near the University illustrates that our efforts at publicity had been eminently successful. This factory made springs for cars, beds and many other uses but not yet for use in space. Owen recounts the experience:

> On appearing at reception, I explained who I was and that I wanted Henderson's to manufacture two matched springs to eject Australis from the launch rocket. Barely believing me, the receptionist asked me to wait while she sought someone to help, or hopefully to eject me. Evidently the man who appeared at least knew of the project and he ushered me into his office, much to the incredulous looks of the receptionist. I explained what was wanted; such and such a size, so much force when compressed, able to withstand the unknown temperatures and so on. Realising that I understood at least something of what I was requesting, he agreed, telling me to return two weeks later when the springs would be ready for me to collect. They were. For me, it was a perfect example of how our publicity had smoothed the way for the project.

One other example is worthwhile telling. Although there was effectively no knowledge of what the radiation environment that Australis would find itself in, if only because we could only guess the orbit, it was reckoned that "space qualified" semiconductors would be needed. By way of simple explanation, silicon transistors (not the field effect transistors of today) rely on charged electrons (and the absence of electrons) that move in the presence of an electric field in a chunk of silicon. The radiation that we were wary of consists of high energy ionising particles, some of which could penetrate well into the electronics of Australis and into transistors. The damage can be temporary, simply leaving a trail of charges that temporarily upset the transistor's operation, or it can permanently damage the very structure of the silicon crystal. "Space qualified" transistors were designed and tested to be more resistant and resilient to both temporary and permanent radiation damage than conventional silicon transistors used elsewhere in domestic electronics, for example. Several requests were met with refusals, not because they did not wish to help but simply because those companies did not manufacture space qualified transistors. However, we eventually contacted Fairchild Semiconductor[4]

Plate 6.9

Plate 6.9 *Launch! What would Occupational Health and Safety people say today about the two crew on the back of the launch truck!*

4   Fairchild was a leader in solid state electronics at the time. However, it was not forward-looking enough for a few who later left to form Intel, the dominant manufacturer of the processor chips for personal computers.

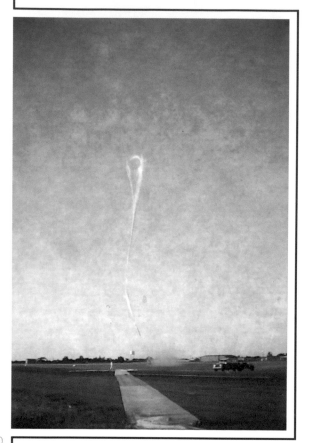

Plate 6.10

in Melbourne who said that they would provide them for us. Fairchild provided all the transistors used in Australis.

Richard and Owen made a trip to Sydney to ask Ducon Condensers for capacitors when, once again, his VW was invaluable. We set off for Sydney on Friday 3 February 1967, at about 7:00am, hoping to reach Sydney that day. Incredible as it will seem to today's young, the Hume Highway connecting Australia's two largest cities had been only recently sealed. It was not uncommon in the 1950s for trucks to be bogged in the creeks and rivers that crossed the dirt portions of the road in New South Wales and for it to be impassable for many days in winter. Imagine the main road between Sydney and Melbourne being blocked! We expected a very long day.

As we travelled up Sydney Road past Brunswick and over Bell St, Coburg, we became aware of a large crowd of people on our right. "What's this?", we asked ourselves. People seemed to be gathered outside Pentridge prison, and then we realised what the gathering was about. Ronald Ryan was about to be hanged. Ryan was convicted for the shooting and killing of a Pentridge warder, George Hodson, while he and fellow prisoner Peter Walker escaped from the gaol, hoping to flee to Brazil, which did not have an extradition treaty with Australia at the time. He was caught in Sydney three weeks later and extradited to Victoria, followed by his trial and sentencing. For the second time in his life Owen had a feeling of dread, this time over the imminent death of a man:

> I was unsure then of the rights and wrongs of capital punishment. (My view now is that it should be abolished, and it has been in Australia.) It was one of those events in life that I will always remember where I was at that moment, others around that time being the Cuban Crisis, President Kennedy's death and the Apollo 11 moon landing.

We drove on rather quietly as we contemplated what had just taken place. And on and on through Victoria and into New South Wales. And on and on through New South Wales until we were flagged down near Gundagai (but not '… nine miles from …') by a woman whose car had evidently broken down. As Owen brought the car to a stop behind hers, a couple of men appeared from behind a bush

Plate 6.11

– this was not quite what it first appeared to be. She wanted fuel and the men seemed to be quite threatening, and so, fearful of what a refusal might bring, we allowed her to siphon fuel from the VW's tank. Unlike women of the time, she seemed to be adept at siphoning petrol, so it occurred to me that this was a regular gig for the three. Once she had enough petrol, there was a request for money, which we fulfilled, and drove away as quickly as we could. Thankfully, the rest of the journey was uneventful and Ducon gave us the capacitors we requested. We returned via the coast – a road less travelled and that adventure was not repeated.

We will delay the visit to the Potter Foundation until the appropriate moment in the story and follow the construction and testing of the satellite itself, the flight model.

Plate 6.10 *Release the payload! The truck manoevres under the 'bubble' so that the payload, when released from the launch truck, does not pendulum into the ground. The payload was released a few seconds later when the balloon had straightened.*

Plate 6.11 *HIBAL headquarters in Mildura in northern Victoria with my trusty VW. The VW was used for the vibration tests - Australis was loaded onto the back seat and driven over rough roads.*

7.1

Plate 7.1 *Paul building Australis.*

CHAPTER 7

How We Built and Tested Australis

By late 1966 we had sufficient confidence in our ability to design and build our satellite and that it would successfully perform in space. Now was the time to build the flight sub-systems, assemble them and prove to ourselves that it performed as expected.

Flight components were being gathered together and delivered to the builders. Yet again, Owen's VW Beetle proved to be invaluable as he journeyed between the homes of Paul, Peter, Dave, Les and Richard delivering components, receiving progress reports, listening to problems they had and goading them on. Owen saw Steve every day as they attended mostly the same lectures and practical classes, and they lived in the same residential college. As Australis progressed, the tempo of conversations with Project OSCAR increased in Les' radio shack. These journeys were not minor trips to the next suburb as Paul and Les lived some distance from Melbourne University and the college where Owen lived, so this became a not insignificant drain on his evening study time during his ever more difficult studies – third-year electronics engineering. It was by no means all lost time for Owen as he was visiting people who practised electronics by building circuits and using them. Owen learned a good deal about the practical side of electronics while studying the increasingly complex theory.

By the middle of third year, Owen was becoming increasingly anxious, even fearful, of the exams at the end of the year and even more so of the final year exams in November of the next year, 1968. He needn't have been but, nevertheless, it drove him to study even harder the difficult subjects of Mathematics – vector calculus, theory of variations and the like – and Physics – Maxwell's equations, relativity and so on. He enjoyed the electronics, especially, and remembers his mother's concern when, in practical classes, he started to work with large electric motors and lethal voltages. Mother was certain that he would electrocute himself or become entangled in a rotating electric motor. Her concern was not without merit, but that was the point, to learn to understand the dangers and to protect oneself against them. If not, the same academic lessons could have been learned with low voltage, low power motors, etc.

Nevertheless, the Australis show had to go on as the sub-systems were completed, tested and assembled onto the Australis structure. Paul did a magnificent and truly professional job with the aluminium frame and structure. By early 1967, Australis was complete and ready to be tested as a whole. One issue that concerned us from the beginning was the manner in which the transmitters interacted (if at all) and certainly how they affected the command receiver. We needed to ensure that everything worked correctly and satisfactorily over the temperature range that we set ourselves. But first, the antennas had to be tuned and Owen has already mentioned the episode with the .22 rifle at Les' home.

Owen Recounts:

> I must pause in the story to describe our living arrangements. Already, we know that Steve and I lived in a residential college at the University of Melbourne. Richard shared a house near the University with Ian and a gentle and genial man who had nothing to do with Australis but who went on to a career with the Bureau. Crucially, however, three very eligible girls, Pauline, Delia and Ronnie, lived in a shared house on the other side of the University. May I say that Richard and Delia were friends? (The plot thickens considerably in due course, but, dear reader, you will have to wait a while for that to be revealed.)
>
> Apparently, Richard had mentioned to Delia that he would like Pauline to 'come around and cook dinner for us one night'. History has not related why he would do that but in any event Pauline offered to cook a meal for Richard's household, a generous offer indeed. As recorded in an interview for a documentary on Australis, Richard had said that they didn't cook much and didn't eat very well. As Pauline remembered, 'the usual sob story, so I said I'll come around ... to cook a roast'. She had already cooked the vegetables before she left her home, the roast was ready to go into the oven, but when she looked in the oven she found 'bits of metal and then this person materialises behind me with Buddy Holly glasses and he says "don't touch that, that's my satellite"'. Richard still wears Buddy Holly glasses. History does, however, relate that dinner was served about midnight, but only after Richard had completed the temperature tests on components of the satellite. It still amuses me to think of the scene, even though, thankfully, I wasn't there.

Despite the initial meeting, the relationship between Richard and Pauline developed as we shall see. Pauline was and is a very talented musician, then as a violinist for the Melbourne Symphony Orchestra.

Now for the temperature tests on the whole satellite. It is true that the army test centre at Maribyrnong could readily undertake the temperature tests, but enquiries showed that they were unavailable at the time we needed them, so we had to be rather more inventive. The temperature tests in Richard's oven that night sufficed for the high temperature testing of Australis.

Low temperature testing turned out to be somewhat easier. Our friend Ian worked in the Department of Glaciology in the University. Departments of Glaciology have freezers that can indeed hold satellites the size of Australis and so one weekend with Ian's help, Australis was fitted with temperature sensors, wrapped in plastic to exclude moisture (we hoped) and set in the freezer. We have pictures of Australis without its outer skins sitting in the freezer, wires attached and surrounded in mist.

Ian continued as a glaciologist eventually working for the Antarctic Division until retirement. Owen met him a few years ago when he was given a flight to the Antarctic as a birthday present. It was Ian who was the resident expert on the flight and they renewed our acquaintance on board. Incidentally, the 12 hour flight was a thoroughly enjoyable experience with passengers and crew relaxed and enjoying the experience. We were invited to the cockpit, where Owen was even asked where he wanted the plane to go. His answer was appropriately Mount Melbourne, an active volcano leisurely puffing smoke into the atmosphere. The pilot pointed to a knob on the dashboard and motioned him to turn it to point the aircraft towards Mount Melbourne. The first and only time he has flown a Boeing 747 aircraft with hundreds of people on board.

All our testing showed to our delight that Australis would survive and operate as we had hoped in space. Frequency drifts with temperature were acceptable, we'd used transistors that were space qualified so really, what could go wrong?

* * *

Around this time, David was experimenting with the University's IBM 7044 computer. His idea was that he might be able to use the computer's "sense lines" to send and receive teleprinter characters between his desk in RAAF Physics and the computer, thereby being able to program it from his desk. An idea before its time. He achieved his goal and Owen remembers him saying that there were several volts of 50 Hz AC mains difference between the computer and his desk. This required some additional electronics to extract the sense line signals from the 50 Hz, but readily done.

Now he could transmit results directly to his desk without having to wait to collect a printout. Mmmmm, this could be useful one day.

* * *

There is a moment in that year that particularly stands out for Richard and Owen. As Australis was nearing completion, it had to be delivered to the OSCAR people in San Francisco and this could be readily achieved by sending it via air. However, we thought, wouldn't it be fun to deliver it ourselves? Once again, we were unconstrained by what was possible, or rather not possible, and so we went about doing it. The question was how to raise the $2,000 per person for the flight. Imagine, $2,000[1] to fly tourist class across the Pacific in 1967. It was

Plate 7.2

---

1   Apparently equivalent to about $22,000 today

Plate 7.2 *Cold testing in the freezer at the Glaciology Department, University of Melbourne.*

Plate 7.3

THE AGE, Thursday, June 1, 1967

OUR FIRST SATELLITE

Three members of the project team (from left) Owen Mace, 21, an engineering student, Richard Tonkin, 24, a law student, and Paul Dunn, 22, an applied science student, with the satellite.

Most dram in GAS h

Lord Casey

an enormous price and utterly beyond the means of all but a very few, certainly not us or our parents. We needed an organisation to give us the money.

We had an introduction to the Ian Potter Foundation, a major philanthropic organisation founded a few years before in 1964 by financier and stockbroker, Sir Ian Potter, who was strongly influenced by Kenneth Myer, son of pre-eminent philanthropist, Sidney Myer,[2] and founder of the Myer stores. The introduction was most probably arranged by Willson Hunter.

However the introduction was made, Richard and Owen visited the office of the chairman on a glorious sunny Melbourne day in late summer. The address, as we recall it, was 95 Collins Street, Melbourne which turned out to be Melbourne's only skyscraper then – of 15 stories or so. Up the lift we went and into the Foundation's reception area. Two rather scruffy university students probably quite inappropriately dressed standing in front of an impeccably dressed receptionist. In due course we were ushered into the office of the chairman, who turned out to be quite a tall and imposing man, also impeccably dressed.

The view from his office was quite staggering for us, who had never been higher than a few stories in any Collins Street building. Before us stretched St. Kilda Road, the brilliant emerald green of the Domain and the Botanic Gardens, the brilliant whiteness of the Shrine and Government House and the blue, blue sky. What a view! This was, of course, the conversation starter for us as we exclaimed the brilliance and beauty of the view laid out in front of us.

As Owen recalls:

> Until, that is, I observed that it was all truly beautiful, except for the then uncovered Flinders Street railyards at our feet below, all dark and dirty. The chairman came over to me, put his arm over my shoulder and said, 'Not to me – I built those rails when I was railway minister in the Victorian government in the 1930s'. This was Sir Robert Menzies, Prime Minister of Australia from 1939 to 1941 and 1949 to 1966, the longest serving of all Australian Prime Ministers, a giant in stature, both physically and in reputation. I reckon I am one of the few people to have

2   The friendship between Potter and Myer led to the founding of the Florey Institute of Neuroscience and Mental Health, a major global player in health sciences.

had the arm of Bob Menzies around me, though whether it was in friendship, understanding or he just wanted to throttle me, I don't know. People either admired him and his policies, or hated them both; there was nothing in between.

I do not remember the immediate reaction to this encounter, either from Menzies, Richard or me; we, nevertheless, explained our case and were dismissed with a promise to look into our request. We were granted two tickets! Somehow, Plessey Components Group were persuaded to give a third for Paul. Perhaps Willson Hunter at work again for us. So three of us were to deliver Australis personally to Project OSCAR in San Francisco. The story of that incredible trip is told in the next chapter.

In the meantime we had to decide who was going to go. No doubt all wanted to go but there were only three tickets. There was generous universal agreement that Richard and Owen should be two but the choice of the third was less clear. At a lunchtime meeting it was decided that it would be put to a vote and that Owen would contact those not at the meeting that day. It is not with any pleasure that he remembers telephoning one team member, explaining what was happening and asking for his nomination for the third seat on the plane. He was less than complimentary about the process and, unfortunately, we have never seen him or heard from him again since that day. Should he read these words, we would be delighted if he would contact us.

It was decided that Richard, Paul and Owen would travel. Peter ruled himself out as he was too busy with his PhD studies, besides it was unlikely that his professor, Vic Hopper, would allow him to go. The dates for travel, departing 1 June 1967 and returning 17 June, covered one week of term break and a week of the second term of Owen's third year:

Could I afford the time? I asked the kindly engineering Professor Moorehouse if I could go. His reply surprised me – 'Yes, of course, you will learn more in those two weeks than we can teach you in a year.' He was correct.

A footnote to the visit to the Potter Foundation. Recently Owen sent a copy of he first edition of this book to Ian Potter's wife with a note belatedly thanking the Foundation for its generosity. She replied in the kindest terms and her letter is now part of the Australis archives.

Plate 7.3 *The Age, 1 June 1967, announcing the departure of Australis for the US. Credit: Fairfax Syndication.*
Plate 7.4 *Tuning a VHF antenna in Les' garden.*

CHAPTER 7 - HOW WE BUILT AND TESTED AUSTRALIS   89

Plate 8.1 *Our Boeing 707 in Honolulu airport 2 June 1967 behind the Japan Airlines 707. Minimal airport security then!*

CHAPTER 8
Delivery

In this chapter we will describe what was for us one of our life experiences, and there were to be two that year, 1967. Richard, Paul and Owen travelled to California, ostensibly to deliver Australis to the Project OSCAR group.

Although the British de Havilland Comet[1] aircraft had ushered in the jet age for long-distance civilian travel in 1952, fatal accidents from the then poorly understood process of metal fatigue prevented it from becoming the aircraft of choice. The Boeing Commercial Airplanes company, benefiting from de Havilland's experience, as well as from its deliveries to the US Air Force, developed the 707, a four-engine jet that has become the model for almost every subsequent passenger aircraft, namely engines slung below and forward of the low, swept-back wings, flying between 9 and 12 km at a little under the speed of sound. The design of the aircraft has to be admired since, despite travelling near the speed of sound, any air flow that exceeds that speed will cause shock waves that add drag to the aircraft. Although the 707 first entered service in the late 1950s, it was not widely patronised on the trans-Pacific route between Australia and the US. This was to be our first international flight.

We set off from Melbourne on 1 June 1967 proudly presenting ourselves at the Pan American (Pan Am) counter at Melbourne's Essendon Airport with our equally proud parents behind us. With the relatively inefficient engines, the flight in our Boeing 707 Jet Clipper was to be long with frequent stops to refuel. Auckland, New Zealand, was our first stop then, we believe, Fiji. Perhaps it was American Samoa. There were so few passengers on board that each of us had three seats on which to sleep across and plenty of rows to choose from, too. As we were approaching Hawaii we were offered Hawaiian cocktails, complete with a purple orchid floating in the martini cocktail glass. Very James Bond! However, not being used to such luxuries, Owen thought it tasted as if the attractive hostesses had raided the aircraft's fuel tanks and mixed in a bit of brake fluid. Shaken not stirred, of course. Richard recalls sampling Drambuie, a whiskey-based liqueur, which he became very fond of during the long flight.

More fuel at Hawaii before the final leg to San Francisco where we were met by OSCAR people who must have been surprised by the youths they saw before them. Owen had turned 21 less than three months before. Nevertheless, we were not abandoned but were taken to our "billets", and what amazing billets they were.

Now, Owen explains that he was thoroughly enamoured by American technology, as you will no doubt observe in what follows. For a young nerdy engineering undergraduate, this experience was scarcely believable. Driving from the airport to Palo Alto where we were to stay, we passed the newly

---

1   Built in Hatfield, England where Owen was to work for a short time some twenty years later.

completed SLAC in Menlo Park. SLAC, the Stanford Linear Accelerator Center[2] was, and remains, the world's longest linear accelerator where atomic and solid state physics, chemistry and biology are studied. Accelerating electrons and positrons up to energies of 50 billion electron volts (50 GeV), it has since provided discoveries of three fundamental particles and been instrumental in four Nobel Prizes. Later, it was the meeting place of pioneers of the home computer and personal computer. Here we were, Australian University students driving past this almost holy place!

Paul and Owen were billeted at Bill Eitel's home in San Jose and, yes, Bill was an amateur radio operator. (We learned to say "San hose-ay" and certainly not "San joe-say" or "San-joze".) He possessed a remarkably large home, what to us was an enormously large number of cars and, as we meandered his garden, a large number of gophers, judging by the holes in his spreading lawns: Owen

2   See https://www6.slac.stanford.edu/

Plate 8.2

Plate 8.2 *A fleet of cars outside Bill Eitel's home. Our Ford Galaxy on the left and his red Corvette with Richard talking to Bill's son.*

had thought gophers were just in the imagination of comic writers, but obviously not. His cars were remarkable for us – not just a Cadillac and a motor home but also a red Corvette! Once inside, Paul and Owen met his wife and were shown to our bedrooms. Owen's overlooked the gophers. It turned out that Bill's neighbour, invisible through the trees, was Judy Garland.

What was truly remarkable was not so much his cellar that seemed to occupy the entire footprint of his large home, and more, but that it was filled with all manner of lathes, drill presses, mills and other machine tools, the sort of things you might find in a well-equipped machine shop. We were to learn later that he had developed a major patent in this workshop! It turned out that Bill was a co founder of EIMAC, a company that made high power vacuum tubes.[3] Bill Eitel and Bill McCullough[4] were working for a manufacturer of vacuum tubes when, in 1934, they formed their own company, which grew to 1,800 employees that produced 4,000 radar tubes *per day* for the US government during World War II. Following the successful bouncing of a radio signal off the moon using "a giant EIMAC klystron",[5] EIMAC became the largest independent manufacturer of high-power vacuum tubes. The company had recently taken over another manufacturer of high-power tubes, Varian. (No anti competition laws then.) We were to meet the two Bills in their shared office later.

What was the technology developed in his cellar? It was a method for reliably joining glass to metal (or was it ceramic?) so that vacuum tubes and klystrons could operate at higher temperatures and hence higher power. This was probably one of the factors that propelled EIMAC to its prominent position in the industry. In fact, EIMAC manufactured cathode ray tubes for televisions that were marketed under the once famous RCA name. Paul was a radio amateur and, of course, knew of EIMAC tubes as they were the preferred final amplifying tube for amateur radio transmitters. Nowadays, many are replaced with transistors. Bill Eitel died in 1989, McCullough in 2001.

Importantly, describing Bill Eitel, *"more than any other single individual, he is responsible for the OSCAR program."* [6]

How little we three knew of the history of wireless technology, amateur radio and OSCAR. How ignorant we must have seemed to the OSCAR people we met as we were taken to places and to meet people of significance to the four OSCAR satellites. As Owen says, he has learned much of this history from researching this book.

---

3   History San Jose, The Perham Collection of Early Electronics' at <www.perhamcollection.historysanjose. org>. Accessed June 2017.

4   In the web page cited in the previous footnote, he is referred to as Jack McCullough. I am sure we were introduced to the two 'Bills' as humour was made of each calling the other, Bill, and further confused when someone entered and asked for Bill. Perhaps Jack came to be used to distinguish between them.

5   A klystron is a special type of vacuum tube used to generate large amounts of power at very high and microwave frequencies.

6   *World Radio* magazine, May 1989, quoted in https://qrqchistory.wordpress.com/about/

Richard was billeted with Bill Hewlett – yes, another Bill and the co-founder of the Hewlett Packard company, which in those days made premier electronic instruments. Bill and his wife lived in a mansion in Palo Alto, and they even had a maid. They made Richard very welcome.

The day after our arrival, we were escorted to Foothills College, the home of OSCAR, by Harley Gabrielson in his Mustang car (a Mustang!). Owen remembers that he understood nothing of the significance of the College to the people of OSCAR, but he certainly ogled at the car.

We were presented with a Ford Galaxy for our use while we were in San Francisco. The fashion for cars with large fins had died and so our Galaxy was "fin-less" but enormous compared with Owen's diminutive VW. Owen has a picture of him at the helm of the car with the speedometer reading 90 – and that is not kilometres per hour!

The next day we were sent off to drive to the natural wonders of the Bay Area, Sausalito, the Big Basin and Muir National Parks, north of the city. Being the only driver, it was Owen's responsibility to get us there – his first experience with such a big car and with automatic transmission, too. To add

Plate 8.3 *Lombard Street, San Francisco.*
Plate 8.4 *Richard and the Galaxy (registration UBG 508) parked in Lombard Street.*

to his confusion, the steering wheel was on the wrong side of the car and – what's this? – the cars drive on the wrong side of the road as well! Richard, the navigator, was given instructions on how to get there via El Camino Real.[7] El Camino Real? Yes, we were told, it was a major road that would take us to the city of San Francisco and it had historic connections with the Spanish missionaries from the late 1600s. While Richard may have been "the natural leader", navigation was not one of his strong points, so it was just as well that Paul had listened well to the instructions.

First, a tour of the city of San Francisco and a drive down Lombard Street, one of the impossibly steep streets on Telegraph Hill, the one with eight tight hairpin bends. We parked the Galaxy below the bends to admire the city (plenty of parks in those days) and the well-planted street. With its soft suspension, the car leaned downhill as Richard appeared to lean toward it in Owen's photograph.

To get to Muir National Park, we had to cross that icon of San Francisco, the Golden Gate Bridge, impossibly high over the waters, and then on to Big Bend where we saw gigantic Redwood trees and tame deer. Another two firsts for us. Then Sausalito where there were floating houses – another first. Back over the Golden Gate Bridge to visit Haight-Ashbury, the centre of the 1960s hippie movement,[8] and what a sight it was for us innocent young Australians. We decided to extend our tour by circling the bay before returning to San Jose via El Camino Real with two more huge bridges and the notorious San Quentin prison *en route*.

On the next day, Monday 5 June, we were taken to the IBM computer plant with a sea of employees' cars

8.5
8.6

7.   Wikipedia, 'El Camino Real (California)' at <https://en.wikipedia.org/wiki/El_Camino_Real_(California)>. Accessed June 2017.

8.   Wikipedia, "Haight Ashbury" at <https://en.wikipedia.org/wiki/Haight_Ashbury>. Accessed June 2017

Plates 8.5 *Famous names on a pedestal of the Stanford University Mills Cross radio astronomy antenna.*
Plates 8.6 *70 metre dish at Stanford's antenna farm.*

in its car park. A memorable moment came for me when we were walking down a corridor and our guide stopped, looked around furtively, and then quickly ushered us into a stark room. It contained a desk with a rather large box, probably a metre cube, with a small, 300mm television screen visible on one side and some sort of keyboard beside it. This, we were told, was the future of computing. What we didn't realise then was that IBM was investigating multi-user, multi-tasking operating systems for its recently introduced IBM 360 line of mainframe computers.[9] These machines were truly pathetic imitations of computers when compared even to today's laptops. The machine on which this book is written is around a thousand times faster and carries a thousand times more memory. Nevertheless, these machines were pioneers in computer architecture. Furthermore, it is possible that the human pioneers of modern computing, IBM Chairman Thomas Watson and chief architect of the System/360 Gene Amdahl, were in the same building as us. Historic, but it takes the wisdom of remoteness in time to realise it as being so. Amdahl went on to found a company that made compatible, and less expensive, computers bearing his name.

Interestingly, specially modified and hardened IBM 360 computers, called 4 Pi avionics computers, were used as fault tolerant computers in the Space Shuttle. As a compliment to their advanced design, the Soviets made clones and several other manufacturers made compatible machines.

Plate 8.7

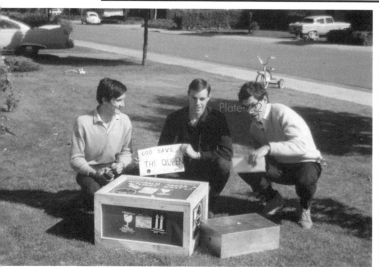

Tuesday saw the arrival of our pride and joy, Australis, so off we went to the airport to collect it and deliver it in its well-padded box to Lance Ginner's home. It had been delayed because of a last minute problem with the command receiver. Nevertheless, here it was and it was opened on Lance's lawn – inside was a message, 'God Save the Queen'. How inappropriate and embarrassing now. Of course, republican sentiment in Australia has become much more popular since then. Oblivious of the future embarrassment, the three of us were photographed with the sign and Australis in its delivery box. Nevertheless, we had accomplished our mission to deliver Australis to Project OSCAR's representative, Lance Ginner. Little did we imagine then that it was to lie in Lance's garage for nearly two years.

The next day, Wednesday 7 June, we were taken to the Stanford University and the SLAC.

Plate 8.7 *Delivery in the US and the embarrassing message. Look at the fins on the car in the top left corner!*

9   Wikipedia, 'IBM System 360' at <https://en.wikipedia.org/wiki/IBM_System/360>. Accessed June 2017.

Thursday saw us visit the Stanford 'antenna farm', a large country area set aside for radio astronomy and antenna development. The largest antenna was a 60m diameter radio astronomy dish. There was even a US Air Force quad helix of similar dimensions, and therefore gain, as ours back in Melbourne. Copycats! It must be said in its praise that it looked a good deal more substantial than ours and no doubt the large building next door to it contained a good deal more equipment than our meagre garret held, and a good deal more sophisticated. Next, we were shown the Mills Cross, an array of relatively small dish antennas perhaps two to three metres in diameter and arranged in a cross. The concrete pedestal on one was graffitied with the names of some intellectual giants who had visited before us – names like astrophysicist Tom Gold, Australian radio astronomers John Bolton, Bernard Mills and A.G. Little[10] (the latter two being inventors of the Mills Cross) among others. We were indeed in august company!

Lance was a young engineer, but older than us, with the Corona spy satellite project, though we knew little of it then. On Friday 9 June, we were invited to dinner with the insanely handsome, crew cut, all American Lance Ginner and his very attractive wife, Wanda, as this was to be our last day in San Francisco. Over dinner, it was announced that we were to go to our next set of meetings in Los Angeles. We had the weekend to drive and Yosemite National Park was a recommended destination.

Saturday morning saw us take possession of another Ford Galaxy. We were told that it was a hire car that had to be returned to Los Angeles, but we wonder whether it was not another example of their incredible generosity. This one took our breaths away … and Owen was to drive it! It had all the features of our previous, yellow Galaxy, plus it was a convertible and it had the unimaginable – air conditioning. Air conditioning in a car! Furthermore it was a

Plate 8.8

Plate 8.8 *OSCAR and Australis teams. From the left: Don Norgaard, Ed Hilton, Lance Ginner, Paul, Owen and Richard.*

---

10   See 'Early History of Radio Astronomy' at <http://www. astro.washington.edu/users/ woody/hra.html>. Accessed June 2017.

screaming bright RED! Oh well, Owen just had to do his duty and drive it to Los Angeles. Off we went, with instructions to present ourselves on Monday morning at … that will be revealed when we get there. We followed the now familiar El Camino Real past NASA Ames Field with its huge dirigible hanger and eastwards to the Napa Valley and Yosemite National Park. While we were surprised at the number of people in the park compared with near zero in comparable Australian parks, it was nothing compared to today's annual four million visitors when it is necessary to reserve a visit!

Yosemite presented itself in all its truly staggering beauty. The iconic images of El Capitan, the redwood trees, the mountain meadows and passes, deer, valleys carved out of solid granite by ancient glaciers, Yosemite River in flood flow were astoundingly beautiful. The effect that it had on Owen was in the number of pictures he took: 42 in all and, unfortunately, all in Ektachrome, not yet colour corrected in their digitised form. Remember that each photograph cost something, unlike today's digital photographs. Forty-two half-frame slides was approaching a full roll of film! Quite an expense for a young undergraduate.

After spending a full day at Yosemite and wishing we had more time to explore the high country, we set off toward our next destination, driving through towns with Spanish-sounding names

Plate 8.9 *The Red Ford Galaxy at Yosemite National Park with Richard photographing Paul who had not seen snow before then.*
Plate 8.10 *Richard standing behind an engineering model of the lunar 'crasher', Ranger.*
Plate 8.11 *The NASA plane that flew us to the Goldstone tracking station.*

like Merced, Madera and Fresno, arriving at the town of Lompoc, surrounded by kilometres of commercial flower farms.

Owen is reminded of two incidents:

> The first occurred when we stopped to fill the car with petrol – no, it's "gas" in the US. The attendant came to the driver's side – it was still the time that attendants pumped petrol, gas, and checked the oil – and said something. I knew it was a question by the upward inflexion at the end of his sentence but I understood not a word. "Excuse me?" I said, and his question was repeated. Now we were visitors and it would have been rude to have asked him to repeat the question for a second time, so I assumed he was asking if I wanted the gas tank to be filled. I nodded and it was.

> The second incident occurred as we were driving in the dark toward our overnight stop. Red and blue flashing lights behind us. Uh oh ... I was not exceeding the 55 miles per hour speed limit so it couldn't be that. Once stopped, the impressive policeman sidled up to the car and asked for my driver's license and I presented his newly minted Victorian licence and an International Driver's Licence, then a necessary accompaniment for foreign drivers.

> "Where y'all from?" he asked.

> "Victoria, Australia" Owen replied.

> "Ah, Victoria, Canada?"

> "No, Victoria, Australia" with the emphasis on Australia.

> He got it this time and repeated, "Ar-stria."

> Having established where we were from, he very politely informed me that one of our tail lights was not operating, "Sir", and we couldn't proceed. Now this presented several problems, the first being a 21-year-old being addressed by a much larger, older and armed policemen as "Sir" and the second being how to get it repaired at night and in a foreign country. He agreed that we would inspect the recalcitrant light and decide what to do. We were foreigners after all. So, Owen got out, lifted the boot (the "trunk"), pushed the light fixture back into place whereupon it lit up satisfactorily and we continued on our way with our newly minted policeman friend wishing us a safe trip back to "Ar-stria".

Richard recalls a third incident, which involved the language barrier. We stopped at a cafe (a diner) in Lompoc for lunch and he asked the waitress for a chocolate milkshake.

"Excuse me?" she asked.

Richard repeated the order, "A chocolate milkshake, please."

"Are you folks from out of town?" she asked.

A fellow customer came to our aid, "Honey," he said to our waitress, "These folks sound like they come from England. The guy wants a chocolate malted, but hold the malt." As Churchill, the World War II British Prime Minister, said, "We are separated by a common language!"

Then on Monday morning we presented ourselves, as instructed, at the gate of the Vandenberg Air Force Base. Now for us space cadets (as Richard is wont to refer to MUAS members), "Vandenberg", as we knew it, was another near-holy place. Its mission was not to fly aeroplanes but to fly rockets and to launch satellites into orbit for the US military and NASA. It was from here that the first three OSCAR satellites were launched.[11]

We announced ourselves to the burly uniform that arrived to escort us into the base explaining where we came from. "Yesirrrr, Ar-stria" came the reply. Oh, never mind. "Leave your cameras in the car", he commanded. That day, we climbed over a launch site for Scout rockets and, I think, the control room. No doubt that thanks are due to our OSCAR friends for the VIP tour of the base.

On to our destination, Los Angeles, where further amazing techie sights awaited us. We were very surprised to see oil wells in the shallows and pumps in gardens idly pumping that oil to refineries that were probably nearby.

On Tuesday 13 June, we presented ourselves (and our showy red Ford Galaxy convertible with its air conditioning) at the Jet Propulsion Laboratory,[12] another near-holy site for us. JPL, as it is universally known, has had a very long history in rocket development, beginning in the 1930s as the Guggenheim Aeronautical Laboratory. During the war years it was operated by the California Institute of Technology for the US Army and nowadays for NASA. Famous names, such as Theodore von Kármán and Werhner von Braun, and famous

Plate 8.12

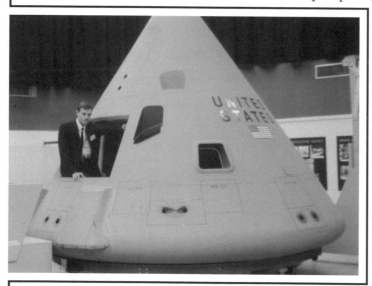

Plate 8.12 *Astronaut Paul checking out the Apollo Command Module for size.*

11   The fourth OSCAR, intended for a geostationary orbit, was launched from Cape Canaveral, Florida

12   Wikipedia, 'Jet Propulsion Laboratory' at <https://www.en.wiki.org/Jet_Propulsion_Laboratory>. Accessed June 2017.

rockets are associated with JPL, including America's first satellite, Explorer 1. More recently, many spacecraft, especially lunar and planetary explorers, have been developed at JPL.

We were met by the likeable Ken Hornbrokk, who showed us around their hall containing models of legendary spacecraft, Surveyors, Rangers[13] and Lunar Orbiter displayed in front of us. In another room a Lunar Orbiter spacecraft[14] was being prepared for flight. We were in the presence of a real spacecraft, one that would be sent to orbit the moon in preparation for the Apollo landings!

There followed on Wednesday 14 June a visit to North American Aviation[15] which later became part of Rockwell International and then Boeing. It was founded in the late 1920s as a company trading interests in the early airlines and in the 1930s it began to manufacture training aircraft. During World War II it produced the Mustang fighter and the Mitchell bomber, among many others, and later such iconic aircraft as the Sabre and Super Sabre of the Korean War as well as the Mach Three[16] Valkyrie aircraft.[17] Interestingly, other divisions of the company researched and built nuclear reactors and defence electronics. It was also heavily involved in the manned space programme which was what we were there to see. North American had won contracts to build the Command Module, the Service Module and the second stage of the enormous Saturn V rocket that carried the Apollo programme to the moon.

Recall that a few months before our visit, the disastrous Apollo 1 fire killed three astronauts, adding yet further pressure to President Kennedy's December 1969 deadline to land men on the moon. Nevertheless, we were led into a gigantic hall where Apollo command modules were being built. On the left, with assembly having been recently begun, was the Command Module for the first manned flight, Apollo 7, that was launched on 11 October 1968. Following a walk through the control room used to test the Command Modules, we were taken to an engineering mock-up of a Command Module and each of us invited in turn to climb into it. Each of us had our picture taken entering it.

In the midst of this excitement we were taken to a place of exotica – an Armenian restaurant. Armenian? We had barely heard of the country and certainly had no idea of where Armenia was. This was a time in Australia of "meat and three veg" for dinner every night and exotic alternative menus such as Italian pastas were just starting to be seen in Melbourne. This was a very different experience for us.

---

13   Wikipedia, 'Ranger Program' at <https://en.wikipedia.org/wiki/Ranger_program>. Accessed June 2017.

14   Wikipedia, 'Lunar Orbiter Program', at https://en.wikipedia.org/wiki/Lunar_Orbiter_program>. Accessed June 2017.

15   https://www.en.wiki.org/North_America_Aviation

16   That is, three times the speed of sound, around 3,000 km per hour!

17   Wikipedia, 'North American XB-70 Valkyrie' at <https://en.wikipedia.org/wiki/North_American_XB-70_Valkyrie>. Accessed June 2017.

While researching the Apollo project, Owen came across this quote on the Wiki page[18] under the heading of Science and Engineering Legacy:

> "The Apollo program has been called the greatest technological achievement in human history. ... The flight computer design used in both the Lunar and Command Modules was, along with the Polaris and Minuteman missile systems, the driving force behind early research into integrated circuits (IC). By 1963, Apollo was using 60 percent of the United States' production of ICs ..."

Given our interest in electronics (and later ICs) and technology, as well as their application in space, of course, we were privileged indeed to have seen what we did, even if we did not fully appreciate the full extent of our privilege.

We had been instructed to return to JPL that evening, 14 June, to view the control room during the launch of NASA's Mariner 5 Venus fly-by. This spacecraft was the back-up for Mariner 4 which was launched in November 1964 and flew past Mars obtaining the first close-up pictures of the Martian surface. Following the success of Mariner 4, the back-up was fitted with additional thermal insulation because Venus is closer to the sun than the Earth whereas Mars is further away. Also, the size of the solar panels was reduced (more solar radiation than on a Mars flight) and a TV camera removed, presumably because the Venusian surface is covered by opaque clouds. Mariner 5 flew past Venus on 19 October 1967, successfully completing its mission and finding that the Venusian surface and atmosphere were hotter than expected.

With all the expected drama, the countdown began and proceeded second by second with the questions and answers:

"Atlas First Stage?"

"Go!"

"Agena Second Stage?"

"Go!"

"Mariner?"

"Go!"

"Communications?"

"Go!"

and so on through all the systems. Then, "ten, nine, eight" and so on, interrupted by "Cooling on"

18   Wiki, 'Apollo program' at <https://www.en.wiki.org/Apollo_program>. Accessed June 2017.

and "first stage ignition", then "Zero" and "Liftoff!" The Atlas Agena and Mariner had begun their journey to Venus and yet we heard not a single roar of rocket exhaust nor could we look out a window to see the night lit by the rocket's exhaust tail. What was unusual about this launch, besides being our first, was that it was the first to be conducted remotely. The launch took place from Cape Canaveral, Florida, whereas the control room was in Los Angeles on the other side of the country. What a privilege to witness that particular launch!

The next day, Thursday, saw us arriving at yet another airfield where we were flown in a NASA Beechcraft twin-engined aircraft to the Goldstone Tracking Station in the Mojave Desert. More antennas, including the 70 m Goldstone "Mars" dish used for tracking interplanetary spacecraft like the one that was launched to Venus on the previous evening.

There were two visits on our final day, neither of which permitted cameras. The first was to the Lockheed Corporation where, incredibly, we were shown a room where glass fibre was being wound around a very large mandrel over one metre in diameter. This, it was explained, was the rocket body for Polaris, a submarine launched ballistic missile – nuclear warhead, of course. Did they really know who we were? Some highly important people cleared to the highest levels of national security? No,

just a bunch of foreign undergraduate students. We shown other things that Lockheed staff were working on but we were just incredulous at seeing the Polaris rocket bodies in production.

The other, Hughes Electronics, produced all manner of radio frequency electronic devices for the military, space and other high-end applications. They were experimenting with stabilising satellites by spinning them and we were amused by one that seemed to be wobbling like a drunk. Next, we were led to a hall where Surveyor 4 was in its final stages of preparation for launch. Its purpose was to measure the characteristics of the lunar terrain in preparation for the later manned moon landings. What a thrill for us space junkies! In his enthusiasm, Richard asked if he could actually touch it. Yes, but only with a gloved hand.

Plate 8.13

Plate 8.13 *Astronaut Owen about to depart for the moon!*

Plate 8.14

Plate 8.14 *Owen and Paul in front of a model of the Apollo Service Module.*

Launched a month later, Surveyor 4 travelled flawlessly towards the moon, and yet, some 150 seconds before it was to touchdown, its radio signals were lost and never re-established. It was thought that the rocket slowing it for landing exploded and the mission itself was declared a failure. Richard sometimes wondered if it was his touch that caused the spacecraft to fail to complete its mission. Unlikely, but on the other hand, Richard ...

This was the end of our trip. We had seen some of the most advanced technological achievements of the day and, in Project Apollo, some of the most incredible achievements in management and organisation. For President Kennedy to be able to inspire and galvanise a country such as the US into such a sustained and challenging project can only be admired and treated with wonderment. As Kennedy had said when he spoke at Rice University on 12 September 1962:

"But why, some say, the Moon? Why choose this as our goal? And they may well ask, why climb the highest mountain? Why, 35 years ago, fly the Atlantic? ...

We choose to go to the Moon. We choose to go to the Moon in this decade and do the other things, not because they are easy, but because they are hard; because that goal will serve to organize and measure the best of our energies and skills; because that challenge is one that we are willing to accept, one we are unwilling to postpone, and one we intend to win ..."

Inspiring words of the kind we need today to lift us out of mediocrity and scepticism.

When we look back at what we were shown, things that most Americans would never see and, in many cases, would not even be allowed to see, it is clear that we were treated as very special visitors. The generosity was simply staggering. The arrangements had the imprimatur of some important people. Willson Hunter must have had a hand in it. We will never know.

Departing Los Angeles on Saturday 16 June 1967, our trip home took us via Hawaii, Samoa, Auckland and Melbourne. Owen had missed a week of the second term but, as his Professor had said, he had learned more about engineering, technology and life from this trip than they could teach him in a year. Richard had a wonderful time, but fell further behind in his law course – space was much more exciting. The sad part is that there were but three places on the trip, but many more who would have learned at least as much as us. Now back to our studies – and the excitement of the Apollo flights culminating in the landing on 20 July 1969, just over two years later.

Plate 9.1 *Try launching three balloons at once! This is a test launch in 1979 of a "Super Pressure" balloon system that provided many-day flights. We flew one such instrument around the world in under two weeks, a record duration at the time.*

CHAPTER 9

Wait for Launch, or Would Australis Come Home in a Box?

Now back to our studies and our normal lives. Not quite.

The telephone rang a few weeks after our return at one of our usual lunchtime meetings. As Owen was nearest the phone, he answered it and on the other end of the line was a woman, one who sounded very, very embarrassed and most uncomfortable. She explained that she was from the Department of Supply, the logistics section of the Australian armed services, and that she wanted to ask some questions about our satellite. After a few moments to gather his thoughts following such a strange announcement, Owen reasoned that we had nothing to hide (a jibe at the Department) and would be happy to answer any questions she had. The answers to her questions were nothing more than could be read in the newspapers, but they culminated in what was the clear intent of the call: when is Australis going to be launched? We understood it to be likely in December of that year, 1967, and so that's what Owen said. Thank you, goodbye. In the following discussion, we all thought it was very odd as there was nothing that could not be found by diligently reading the newspapers which is what, after all, my spy father did for the Department of the Army, although his interest was Indonesia.

All was revealed the very next day when WRESAT,[1] a satellite from the government Weapons Research Establishment (WRE),[2] near Adelaide, was announced. That was what the telephone call was about! This should not have been so surprising as Owen had written to his local member of Parliament, Peter Howson, Minister for Air, telling him of our intentions. Owen has his acknowledgement letter dated 4 March 1966 thanking him for his advice to the government.

In 1966 and 1967, the US had been conducting tests from the Woomera test range in South Australia on the re-entry of nuclear warheads from space. Ten Redstone/Sparta rockets had been sent to Australia for the tests, along with a team to launch the rockets. Sparta was a liquid fuel Redstone first-stage rocket with solid fuel second and third stages. Redstone was developed by Werner von Braun as a modern version of the wartime V2 missile and it launched Explorer 1 and Mercury astronauts Alan Shepard and Virgil Grissom.

1   See http://www.honeysucklecreek.net/supply/WRESAT/main.html and
    https://www.dst.defence.gov.au/innovation/wresat-%E2%80%94-weapons-research-establishment-satellite
2   The Weapons Research Establishment (WRE) was a joint project between the UK and Australia for testing and proving military weapons, including rockets and missiles. (Its remit was widened and it became the Defence Science and Technology Organisation, DSTO, and now the Defence Science and Technology Group, DST Group.) The test range was, and is, located in the desert in South Australia and extends into the Northern Territory and Western Australia. The research establishments are located throughout Australia, principally in South Australia.

The flights had proceeded with no failures (Werner knew how to make rockets) and so the spare Sparta was not expected to be needed. Someone suggested that the unused rocket could be used to launch an Australian satellite and the idea took hold. Let's rephrase the last sentence with an eye to the conspiracy theorists and say that, in view of the possibility that Australis could be launched before any official government satellite and much to the government's embarrassment, the unused rocket could be used to remove that embarrassment.

Let's add something else for the conspiracy theorists. An article on the front page of The Australian on 15 April 1966 with a headline, "U.S. to launch our first satellite" including a picture of Peter, Owen, John and Richard holding a model of Australis, would have caught the eye of people in the National Capital. Might that have been enough to push government officials into approving WRESAT?

Needless to say, there is no proof either way but that is the stuff of conspiracy theories after all. Nevertheless, WRESAT was launched on 29 November 1967 conveniently before the expected month of the Australis launch. That a launch of WRESAT before the end of 1967 was part of the agreement with the US is left unsaid by those conspiracy theorists.

NASA had agreed to track WRESAT, no doubt Willson Hunter again, and the US paid for the TRW launch team while the people at WRE set to work. In under a year a complex satellite was built, tested and flown. It extended the work that WRE had been doing in the upper atmosphere with the University of Adelaide, in particular the heat balance between the sun and Earth, a pre-cursor to today's climate modelling. More specifically, detectors measured ultraviolet radiation and X-rays in four wavelength bands to understand the distribution of oxygen and ozone in the upper atmosphere as well as the physical processes taking place.

There were also goals, parallel to ours, of extending scientific, technological and management disciplines.

Generously, the Department of Supply invited us to the launch of WRESAT at Woomera in the South Australian desert. We had to get ourselves to Adelaide and we would be flown to Woomera for the launch. Richard and Peter made the overnight train trip with owen, the smallest of us, sleeping in the overhead luggage rack, but the launch was scrubbed that day and we returned as our booking demanded. WRESAT was launched the next day! It's just a pity that we didn't think to take the VW. (As you will read later, Richard did attend the launch – there are some inconsistencies in our recollections.)

WRESAT operated on its batteries for about five days before diving into the Atlantic Ocean on 10 January 1968 after 642 orbits. At least we can boast that Australis remains in orbit and will do so for 100,000 years more, give or take a few thousand!

Many years later, Dick Smith, the Australian explorer and electronics shop founder, discovered WRESAT's first stage in the South Australian desert. It was recovered and is now exhibited in the museum at Woomera. It was the last flight of a Redstone rocket.

* * *

Yet 1967 was an even more significant year for Owen and, once again, the VW figured in events. Richard suggested a ski trip to Mount Buller, a snow field north of Melbourne and we would stay in the University Hut on the mountain. Now Owen was an experienced skier, so he jumped at the opportunity which came with a condition – he was to drive the team to the ski field. And, it turned out, the team was to consist of Richard, Owen and the three girls from Parkville: Pauline, Delia and their friend Ronnie. We were not going to fit five people in the VW without some considerable squeezing, but that was mere detail. As it turned out, Richard came down with a cold so the three girls and Owen set off for a weekend skiing, all squeezed into the VW beetle.

None of the girls were experienced skiers so it fell to Owen to help them. Pauline and Ronnie had spent an earlier weekend with Richard and a friend Rex, so at least they had a couple of days on the snow and knew what to expect. (Owen probably should not say that they were then notionally their respective girlfriends and boyfriends, so he won't.)

Delia had never even seen snow, so it was a totally new experience for her. Needless to say, the gallant knight attended her on the beginner's run, Bourke Street, until she fell awkwardly and decided to return to the hut, leaving Owen to try his hand on the Bull Run for experienced skiers. It was late in the season, snow sloppy, so it was not much of a day.

Richard's arrangements for us in the hut were … unusual, to say the least? Perhaps not for the 60s! We all slept in a bunk room and the wicked Delia insisted that the only male in the room kiss each of the girls good night. Perhaps Owen was mistaken, but it seemed to him that Delia's kiss was a little more passionate than the others.

Delia had fallen badly and her leg was hurting her so, given the less than perfect conditions, we began our return journey around midday on Sunday, calling in at the Mansfield Hospital where her leg was x-rayed. A small fracture.

Owen proposed to Delia in the early hours of New Year's Day, 1968, outside his family's home in Melbourne. She accepted and they were married on 11 April 1968. Owen says that was the best decision of his life. He and Delia recently celebrated their 50th wedding anniversary.

* * *

Plate 9.2

Plate 9.2 *Owen's electronics designs were becoming more complex! This is the electronics package for an instrument to measure neutrons emanating from the sun.*

Diverted as he was by his blossoming romance, study called and weather satellite pictures had to be received while we waited for word from OSCAR on when Australis would be launched. And waited. And waited. It seemed that the US Air Force had lost their enthusiasm for flying some foreign satellite build by a bunch of school kids – no university kids – and so Australis languished and languished in Lance Ginner's garage for two years.

Owen recalls that, as the 1968 University year got under way, Australis slipped further from his mind. Complex mathematics, fascinating electronics and preparations for the wedding came to occupy him.

It was Owen's family's custom to go to a restaurant for lunch each Sunday and by that time Delia was regarded as a member of the family. On the Sunday after Owen's twenty second birthday he and Delia arrived (in the VW, of course) at his family home to discover that his father had a very bad headache and we wouldn't be going to lunch. X-rays showed that it was an aneurysm[3] deep in his brain and nothing could be done for him. He died on 10 April just three weeks before the planned wedding day.

Showing remarkable leadership, Delia arranged for the wedding to take place the next day with only the mothers and sisters attending. (Delia's father had died the year before under very similar circumstances, with the same consulting doctor and in the same hospital room.) They moved into Owen's parent's home as they had planned, as his parents had been due to travel to England for him to see his brothers and their families, but it did not happen. Owen records the actions of the Trinity College chaplain, Rev. Barry Marshall, who married him and Delia despite the traditional ban on all ceremonies after the altar had been stripped for Easter.

Life changed with early married life and as final exams approached. Owen passed with Honours (not a bad achievement, given all of the above) and was left with a decision on what to do next. A year south with the Antarctic Division was out of the question now and Boeing Aircraft was no longer hiring Australian graduate engineers. By no means a last resort, Owen chose to join Professor Vic Hopper's department and study for a PhD in balloon-borne astronomy.

As previously mentioned, Vic was interested in balloons as a means of reaching above the Earth. He had formed a partnership with Professor Glenn Frye at the Case Western Reserve University (Case) to study gamma ray astronomy. As gamma rays are absorbed by a remarkably small amount of air, it is necessary to get above a large part of the atmosphere to measure gamma rays from a stellar origin. Balloons.

Owen was assigned (you were assigned in those days, at least in Vic Hopper's department) to study under Dr. John Thomas, whose interest was infrared astronomy. Readers will not be surprised to learn

---

3   A brain aneurysm is a ballooning of a blood vessel in the brain and in Owen's father's case it burst, eventually killing him.

Plate 9.3 (Opposite page) *Early morning launch of our gamma ray telescope from Palestine, Texas, December 1978.*

that the wavelengths that John was interested in were absorbed by the atmosphere requiring telescopes to be carried above much of the atmosphere. Balloons.

Large unmanned scientific balloons, of the kind we had tested Australis sub-systems on, were to be Owen's life, and Delia's, too, as a balloon widow, for many years to come.

As Owen was an electronic engineer, no matter how inexperienced, he became the department's engineer and so was set design tasks in a very wide range of fields. He was a generalist with no specialist skills in any particular area of electronics. He does, however, describe two interesting electronic exploits , among many.

Plate 9.3

Initial flights of the joint Case-Melbourne gamma ray instrument, a 2x1x1 metre, 200 kg machine showed that any gamma ray sources were too weak to be detected by the relatively small (300x300 mm) spark chamber detector. However, it was possible that a few pulsars[4] would be detectable but this required that measurement of the arrival time of every potential gamma ray event to an accuracy of one millisecond, and preferably much better. This on a balloon in a near-space environment with 10,000 volt sparks being set off routinely (this was part of the detector) and hundreds of kilometres from the launch site, often out of radio range. A system was devised but it did require synchronising with standard time (Greenwich Mean Time). This was less than easy when the launch site was in Alice Springs in the centre of Australia.

There was one spark of hope – the super secret Pine Gap satellite receiving station was nearby. Owen's team managed to persuade a staff member to broadcast a timing signal, illegally, from Pine Gap. By the time the signal reached the receiver at the Alice Springs Airport, it was very weak, barely above the background noise. It took hours of careful scanning to find the very narrow pulse but once they had, they could synchronise the on-board clock and hence measure the time of arrival of the gamma rays to microsecond accuracy. This led to extending the detection of pulsed radiation from the Crab and Vela

4   Pulsars are stars that emit electromagnetic radiation, over an extraordinarily wide range, in highly periodic pulses.

pulsars by nearly a thousand-fold, from high x-ray energies (50 MeV) to medium gamma ray, perhaps 10 to 20 GeV. Pulsars are stars that emit electromagnetic radiation, over an extraordinarily wide range, in highly periodic pulses.[5]

It is worth recording that among the more than a dozen balloon flights that Owen was involved with, his group held two records – the highest flight of 156,000 feet (about 47km) and the longest of 12 days. No doubt both have long since been broken.

Later, while doing his post-doctoral year with Glenn Frye and his team at Case in Cleveland, Ohio, Owen was presented with the problem of packing many thousands of electronic detectors of what were essentially arrays of Geiger counters into a tiny space. The obvious answer was to have custom integrated electronics built, but this was well beyond their budget, so Owen devised a much simpler scheme based on available microelectronic chips. The idea, he was told, was later used on the NASA Compton Gamma Ray satellite which carried a much bigger detector that greatly extended the number of known gamma ray sources.

Incidentally, Case was the site of the famous Michelson-Morley experiment that led to Einstein's Theory of Special Relativity; that the speed of light is constant, irrespective of the speed that the observer is travelling or the direction she is looking.

* * *

So, Owen's life continued with Australis fading into the background.

Plate 9.4

5   The energy of atomic particles are generally measured in electron volts (eV), so that MeV is the unit for a million electron volts and GeV for a billion electron volts (eV). A 50 GeV gamma ray has the energy of an electron accelerated by an electric potential of 50 billion volts. One eV is approximately 1.6×10^{-19} joules; not a lot for us but significant for an electron!

Plate 9.4 *The improved gamma ray detector, a multi-wire proportional counter. There are over 500 of my "electronic detectors" visible in the image (but very hard to see) on each of the eight trays of proportional counters. Each tray was about one metre square.*

By mid-1969 he was in the midst of his PhD studies and a baby was on the way, due in January or February 1970.

Plate 9.5

Owen remembers a singular day, one of those that one remembers forever. After the setback of a fire, the Apollo 1 capsule in 1967 in which three astronauts were killed, Apollo 11 was to be the first manned landing on the moon. It seemed that everyone in the world was glued to television sets to watch the launch on 16 July, land on 20 July, and then, at home for lunch, Delia and Owen watched Neil Armstrong step down off the spidery lunar lander and onto the surface. Buzz Aldrin followed him twenty minutes later. Despite the poor black-and-white quality video, which improved markedly with later flights, the moment was truly memorable and historic. Each successive landing produced better video quality and the final flight, Apollo 17, even showed the lift-off of the Lunar Module's return to lunar orbit stage *in colour!*

It is interesting to learn that the origins of the Apollo programme began during President Eisenhower's administration.[6] It was Soviet Russia's space feats, including the manned space flight of Yuri Gargarin on 12 April 1961,[7] and political pressure to ensure American superiority over the Soviet Union that spurred the new President, John Fitzgerald Kennedy, finally to commit the US to its manned lunar programme.

Memories of Australis receded until ...

6   Wikipedia, Apollo progeam' at <https://en.vikipedia.org/wiki/Apollo_program>. Accessed June 2017.
7   Wikipedia, 'Yuri Gargarin' at <https://en.wikipedia.org/wiki/Yuri_Gargarin>. Accessed June2017.

Plate 9.5 *Apollo 11 moon landing as seen on our small black and white television at home. One astronaut is clearly visible in the foreground of this fuzzy image, another on the left in front of the lunar lander.*

Plate 10.1 *Australis OSCAR A underging testing at NASA Goddard. From the left the HF transmitter, space for one of the ejection springs, the HI keyer and the VHF transmitter are visible with the battery packs behind.*

CHAPTER 10

The Flight

Recall that Jan King was a member of the board of directors of AMSAT Incorporated, the east coast group who were interested in continuing the work of Project OSCAR once it became clear that US Air Force launches from the Vandenberg Air Force Base were most unlikely to be repeated. This group had senior technical and managerial people, as well as the junior engineer, Jan King, who was given the task of preparing Australis for launch. All were radio amateurs and all had day jobs with NASA, companies and universities.

While there was a German project progressing, and it became OSCAR 7, they realised that Australis might be their entry into amateur radio satellites. As Jan recalls, "We knew only the general story because it had been reported (in the general and amateur radio press). Remember that Owen and Richard had done a very good job of publicising what they'd done. They were on radio and it had been in any number of articles, including the major amateur radio magazine called QST." Thanks, Jan.

It made sense for them to take over Australis from Project OSCAR. And so it was shipped to Washington in late 1968 or early 1969.

Jan continued, "Our impression of the spacecraft itself was – I was looking at a lot of flight hardware every day – considered to be all right, although the (outside) case didn't quite look up to the job. However, when we managed to turn it on and operate it we realised that a lot of work had actually gone into it … We did refurbish a number of things (including the batteries) but it's fair to say that the primary sub-systems they'd developed were flown pretty much as is. We had to do a couple of things we found in testing where something wasn't quite right and had to change a couple of components for better quality as they didn't have access to the kind of parts that were sitting in a drawer downstairs and I could just walk down, sign a little piece of paper and take the hardware. That was all I had to do to get the parts." We understand that additional filtering was required on the transmitters.

It is well worth recalling what was happening in 1969 when Jan and AMSAT were testing Australis in preparation for a launch. Apollo 11 had landed on the moon in July with the ever memorable remark, "That's one small step for a man, one giant leap for mankind", and Apollo 12 in November. It was not yet time for the disastrous explosion of Apollo 13 in April 1970. NASA was completely taken up with the Apollo program and then there was this annoyance of a foreign satellite wanting a piggyback ride on a NASA flight.

Following filings with the Federal Communications Commission (FCC), there were, as Jan describes it, two political problems and a technical one. Jan was more than capable of handling the technical issues we've already described – he was a radio amateur, after all.

POLITICAL ISSUE NUMBER ONE.

Australis was, of course, not an American satellite and there was no precedent for flying foreign secondary payloads on US government missions. Furthermore, this Australis satellite had been built by non-professionals with no experience in building spacecraft. However, Americans were working with the Australians, so there's professional American involvement.

By that stage, NASA had flown secondary payloads to test the Apollo network, so there was precedent for flying secondary payloads and, besides, the US Air Force had flown the four OSCAR satellites as secondary payloads. So, if the Air Force can do it, so can NASA. Besides, it was an international programme with a friendly and supportive country that trained young people "so there's education involved in the process". They had a good story for NASA decision makers. Political Problem number one pretty much solved, so now for ….

POLITICAL ISSUE NUMBER TWO.

Who owned the flight and, thus, who approves flying the secondary payload?

Now, this is not as straightforward as it might seem. The proposed flight was with TIROS M, which was a programme of the National Oceanographic and Atmospheric Administration, NOAA, but the launch rocket was owned by NASA. As Jan succinctly put it, "They (NOAA) don't own the rocket, they own the ride" and the spacecraft. NASA owned the rocket. So, who approves a secondary payload? This was a new problem that had not been encountered before – recall that the US Air Force owned both the satellites and the launch vehicles, so they had complete control over their flights.

At that moment, NOAA was transitioning their weather satellites from the development of a system to help weather forecasters to an operational system. TIROS M was to be a proof of concept for the operational system and so Australis was to fly with the first operational satellite and an improved weather satellite system.

To further complicate things, the NOAA satellite was to be launched on a Delta rocket. The Delta was to become the standard launch vehicle for government satellites. The Delta project was continually improving performance and the vehicle for the TIROS launch had greater performance, and greater complexity and risk, than was necessary for the TIROS flight. Then add the risk of a foreign secondary satellite. Using AMSAT's board's connections, a presentation was made to the satellite managers and at the end of the hour and a half, the response from NOAA was, "I have three responses to your presentation in terms of how NOAA feels about this proposed secondary mission … no, hell no and never. You can take your pick from those." Most people would interpret that as negative!

However, Jan was encouraged by the smile of a Delta rocket person (John Tomasello) in the room. Once again the amateur radio mafia was called upon to introduce him to the NASA Deputy Director for science, Dr. Dick Marsden. Yes, Marsden was an amateur. Marsden decreed that Australis was a science mission and, therefore, he had jurisdiction. Furthermore, he said yes, NASA would fly it on the

Delta. There was one provision, however. NASA would pick up the tab! No cost to AMSAT!

An AMSAT member, Dick Daniels, worked in the NASA headquarters and so he was able to place Marsden's letter of approval before Thomas Paine, NASA's Administrator, who signed it. By the time Paine had signed the letter approving the secondary payload, the Delta project was already preparing hardware to mount Australis on the second stage.

One of Paine's stipulations was that they had to get permission from the Deputy Associate Administrator of NOAA, Jack Townsend. That presented few problems since Townsend was … yes, a radio amateur. However, Townsend required an entire environmental test programme, which was undertaken – mostly.

Now for the Technical Stuff

Once the political issues were overcome, work could begin on technical matters. Jan recounts what he and the team did:

> We began to test it and understand it and then to modify some of the items that needed to be adjusted. We reviewed the magnetics for the system and measured them.

> We worked with the Thermal Design Branch at Goddard who assigned one of their junior engineers to the project. He didn't quite get it right but, installing the thermal control coatings on the satellite was the most transformative thing that happen to the spacecraft. That was done in the Thermal Control Lab at Goddard. It went in as a rather ordinary looking rectangular box and came out looking a beauty with its shiny chrome surface and two thermal blankets (one on either side where the 10 metre dipole sections were mounted). Another thing most have missed: the black surface you see in most famous final photo (Plate 10.6) makes most of the surface look black. That black stuff is called "strip coating" (a rubbery material) and it was to protect the sensitive low emissive surface (chrome) that was below it. Just before we mounted the S/C to the launch vehicle this was stripped off to reveal the very bright surface underneath. The black coating was never part of the final solution.

> The design was not quite right. We needed a surface with a higher emissivity and a lower absorbivity, as the temperatures were about 20C too high. So, we should have added a bit of white paint or a material called OSR (optical solar reflector) to a fraction of the outer surface. But this was my very first satellite and you learn by making mistakes.

> I learned a heck of a lot on this mission but, Australis was a part of my training, sad to say. Still we made it work but, it would have worked for a longer duration if we had gotten the thermal design right. In the case of the thermal design, it was done by a guy who became a friend for many years. But I virtually memorized his equations, I still have the book he used to do this analysis and I still occasionally use it to remind myself today of how to do it.

So, this was not a "casual" experience. Ted Mikalak, the young thermal engineer, remained a friend for many years until I left Goddard in the 1980s. Our thermal designs got better over the years in AMSAT as Dick Jensson (WD4FAB) and I began to do this by ourselves, without the help from NASA.

Jan describes the laboratory in the basement of the Goddard building where Australis was refurbished and prepared for flight by a group of "bright young engineers." Their managers gave them the freedom to do what they wanted to do, realising that "it wasn't exactly in line with our position description at work. This was teaching us something that we couldn't get from everyday engineering," recalls Jan.

Among the "everyday engineering" that Jan was employed to do was testing one of the lunar rovers used by the astronauts and seen by countless millions around the world as they drove (or perhaps hooned) around the moon's surface. At the time, there were 15 spacecraft in the building being tested, all multi-million-dollar missions, and here they were playing with this tiny Australis project.

One of the obvious paradoxes between Jan's experience at NASA Goddard and ours in Melbourne was the enormity of the difference in resources of every kind; knowledge, experience, intellectual and physical. The only resource we had other than our own knowledge gained through study and amateur radio was the University's Bailleau library. As the reader can imagine, less than a decade after the launch of the first satellite, there was precious little in the way of technical literature in the library. Especially so, since at the time the Soviet Union and the US were in the midst of the Cold War and almost any information about rockets and satellites was considered to be of military significance.

An obvious modification that was performed was to apply a thermal layer to the outside skin of Australis. When built, we knew that some sort of surface treatment would be necessary to control Australis' internal temperature but had no idea whatsoever of what this might look like. We painted the Australis model we used for publicity in black-and-white stripes, not to recognise any allegiances to a Melbourne football team but to mimic OSCARs 1 and 2. The surface applied at Goddard was a black rubber compound over much of the surface area with a square of gold foil around the antennas giving the appearance familiar to us now. Just before launch the rubber compound was removed to reveal a bright chrome surface.

Jan mentions that he is unsure whether the springs were replaced. We can't think why they would need to be, so we will say, tongue in cheek, that because Australis OSCAR 5 (as it is known once in space) entered orbit with a relatively low spin, they must have been the Australian springs!

* * *

It's worth mentioning that Owen has a distinct memory of being told, perhaps by Willson Hunter, that approval for the Australis launch had to be obtained from the political head of NASA, the Vice President who in 1969 was Spiro Agnew. Nobody else confirms that memory.

* * *

Once we had word of the possibility of a launch and later as it became more certain, MUAS began to mobilise. Owen dusted off his antenna pointing prediction computer program, which he called Standard Orbits, and began to write a Users' Guide.

In 1969 there were no copying machines as we know them today. No word processors, only typists, mainly women, so his first task was to persuade Maeve, Vic Hopper's secretary and the Department's typist. Yes, she would type the Users' Guide for me without charge but we had to be careful because, recall, Vic Hopper wasn't favourably disposed to this "silly satellite stuff". Next, we wanted some pictures and calibration graphs, so a small number of lithographic plates were donated.

The frontispiece carried pictures of the striped Australis and a view of its internals. The first page of the Guide, which is reproduced in Appendix A, carried the words:

> "The Australis OSCAR A Users' Guide contains full instructions for all wishing to track the satellite. Since the success of the project depends on the support of a large number of tracking stations, we are anxious to enlist the co-operation of suitably equipped radio amateurs, shortwave listeners and VHF enthusiasts everywhere."

The opening paragraph of the first chapter of the Users' Guide, Background, is interesting to view in the light of today's knowledge:

> "The Melbourne University Astronautical Society was formed at a time when the image of space research was dominated by a spirit of adventure. Today, much of the popular interest has subsided, but the potential of spacecraft is being rapidly revealed. The satellite is an indispensable tool in many fields of research; its use in communications, navigation and meteorology is commonplace."

This, with the Apollo moon landings not far away!

The second chapter described each of the sub-systems in a similar manner, but less detailed, with block diagrams of the electronics and a drawing showing the configuration of each of the subsystems within the satellite.

Chapter Three contained a description of the organisation of Regional Directors for each of the three International Telecommunications Union regions and Local Co-ordinators for Asia and Australia, including the Australian states, Japan, Malaysia and New Zealand. Next followed instructions for the co-ordinators to provide antenna pointing directions to users within their region. There included a sample print-out from the IBM 7044 computer – Owen was concerned that he might be reprimanded for printing off hundreds of pages of the same figures that were part of the Users' Guide. He was not, perhaps because of certain blind eyes or perhaps the print-out was not noticed.

The final chapter showed how to receive the telemetry signal and how to use the calibration graphs provided in the Guide to determine battery voltage and current, temperatures and the satellite spin. A sample completed coding form was included.

Finally, there was a page of acknowledgements of donors.

Several hundred Users' Guides were roneo'd[1] and distributed to amateurs via the coordinators. It is reproduced in Appendix A.

* * *

The press was alerted and articles published in magazines to ensure that as many amateurs as possible were excited and prepared for the flight.

* * *

Some weeks before the launch Owen was telling his wife, Delia, about the satellite and, seeing his excitement, she volunteered that she would like to see where all this happened, that is, our meeting room. It should be mentioned that she was very near to term with their son, Nicholas. Off they went one lunchtime to the University. (Thankfully by this time Owen was a very junior staff member and had a car park nearby.) Into the Natural Philosophy Building, up the stairs, out onto the roof to be presented with the near-vertical ladder to our meeting room. Heavily pregnant, Delia was not put off by this obstacle and she bravely struggled

Plate 10.2

1   A relatively cheap but antiquated and dirty means of producing multiple copies of a document from a master sheet. that had been cut by a typewriter. The typist had to hit the keys heavily to ensure that the master was cut through so that the printing ink could flow through it to the paper. Considering the technology it produced a remarkably good quality product.

Plate 10.2 *The command receiver being modified.*

up the ladder and into the tiny room, much to the surprise of those inside. Nevertheless, she saw something of what the excitement was about before climbing down the ladder and returning home.

Nicholas was launched less than two weeks after Australis. Not good timing for a newly arrived baby and a satellite!

* * *

As the launch approached, a regular "sked" on amateur radio was established, we at Les Jenkins' home and Jan and his team in Washington. The launch window narrowed as we got closer to launch and the initial orbit parameters for TIROS were made available. Off Owen went and eagerly ran his Fortran program again to produce the "Standard Orbits" for TIROS and, hence, for our satellite. Once the launch date became fixed, the information was disseminated.

Let us step back for a moment from the preparations that we were making in Melbourne. As Jan acknowledged, our publicity had been a major part of the Australis project and we were not going to let it slip at this point. We wanted to invite the press, contributors and friends to the launch, as impractical as it may sound. However, if we could re-broadcast a commentary from the control room of the Delta-TIROS launch into a room in the University, that would have to suffice. In those far-off days, interstate telephone calls were expensive and international calls were simply unaffordable. So, once again off to the telephone system provider, in those days the Post Master General's Department (PMG), a federal government monopoly of postal and telephone services.

We have no recollection of how we managed to contact the appropriate people in the PMG, although Owen had done a Christmas holiday work placement in the Department so perhaps he used his contact there. Steve was working for the PMG by then, so he may have made the contact. Perhaps it was Willson Hunter again. However it was done, we managed to persuade the PMG to give us a line between Washington and Melbourne for one hour on the evening of the launch. The normal charge would have been $600 for that hour!

Back to Washington for the launch. Jan and his AMSAT colleagues were invited to the Mission Director's Centre in Washington for the launch. He recalls the moment that our satellite separated from the Delta rocket and it became Australis OSCAR 5. They were intently watching a strip chart recorder[2] when the pen moved producing a blip on the paper – this was the moment of separation. The chief technical officer at NASA congratulated Jan. What a moment that must have been for a relatively junior engineer who had been moving in the highest circles of the organisation.

Back to Melbourne where the telephone line in our meeting room garret was extended into the lecture <u>theatre below</u> and an amplifier rigged up to broadcast the commentary into the room. (That an ordinary

2   A strip chart recorder is a machine with a steadily moving roll of paper with a motor driven pen that writes on the paper strip, thereby recording changes to an electrical signal, for example.

person dared to meddle with the PMG's telephone lines and actually broadcast the conversation was undoubtedly illegal. Never mind). Owen had prepared a large world map and sketched the first few orbits onto it, highlighting it so that everyone in the theatre could see the track. The telephone connection was made perhaps half an hour before the launch as the press, friends and members gathered for the launch itself.

The drama and tension escalated as the countdown proceeded.

"First Stage?"

"Go."

"Second Stage?"

"Go."

"TIROS?"

"Go."

Then the countdown with everyone in the room holding their breath. Fingers crossed, hoping.

"… three, two, one" an eternity waiting, then "liftoff". Our baby was on her way into space! It was 9:37pm on Friday 23 January 1970, in Melbourne, a moment few of us in MUAS will forget.

Jan doesn't recall much of the launch except to say he was about as excited as he had ever been before, having actually watched this thing disappear into the sky via live TV from the launch site. Delta allowed him to be in the Mission Director's Center for the launch. He was thousands of miles away at the time of the launch but, he got to watch it from a real control room - an absolute dream come true. We were pretty excited, to say the least.

Jan remembers the cheers that erupted when the TIROS satellite separated from the second stage Delta rocket. We heard those cheers in Melbourne and the later cheers as Australis was gently pushed into its own orbit by the Henderson springs, the micro-switches operating and connecting the batteries to the electronics. She was now to be known as Australis OSCAR 5, or AO5.

The launch took place from the Vandenberg Air Force Base in California that we had visited two and-a-half years previously and the launch was near perfect, as close to the design orbit as it could be. We traced AO5's path from Vandenberg on the west coast of the US southwards over the Pacific Ocean, over the top of Mururoa Atoll where the French had conducted their nuclear weapons tests, toward Antarctica then northwards between Madagascar and the African continent. Crossing the equator, she flew along Italy, over Norway and Greenland, before turning south again and passing over Canada, the

Cook Islands and onwards. It was not for some time that we learned that AO5 had been heard soon after separation by a radio amateur in Madagascar.

One-and-a-half hours after the beginning of the telephone call we hung up, although the PMG people did not seem to be in a hurry to finish the call, even if the cost would now have been around $900!

At the same time, Les had been monitoring AMSAT broadcasts from his radio shack in Cheltenham. It was not until the second orbit that a New Zealand amateur heard AO5 and some three-and-a-half hours after launch that the signal was picked up in Australia. It was a weak signal at a low elevation as AO5 passed southward down the Tasman Sea between New Zealand and Australia. The next orbit passed southward over Australia, over Townsville, Queensland, and Mount Gambia, South

Plate 10.3

Plate 10.3 *The command receiver being modified.*

Australia, some five hours after launch. A few dedicated amateurs heard AO5 at around 2:30 am on Saturday morning.

* * *

Here is Rod's recollection of the launch:

I was a PhD student when the satellite was launched, and that year I was President of the Astronautical Society. We invited a contingent of reporters from the local press to a lecture theatre in the Physics Building, close to the Astronautical Society clubrooms. A telephone hook-up to the launch site was organised and the countdown was relayed over the phone.

At our end of the phone hook-up, Richard communicated with the launch site and provided his own commentary. We tapped into the phone line and played the commentary from the US, together with Richard's commentary over the public address system in the lecture theatre. Unfortunately, Richard's voice was much louder over the public address system than the weak voice from the US. We had to ask Richard to talk quietly, so that we could adjust the volume to a level that enabled the voice from the US to be heard without Richard's voice dominating. But as the launch proceeded, Richard became more and more excited and voluble. He more or less completely drowned out the commentary from the US. It was fantastic to watch Richard's excitement (which we all shared). There was no way to quieten him down!

* * *

And here is Richard's recollection of the launch:

Some two years after Peter Hammer and I laboriously tapped out our letter to Project OSCAR on someone's manual typewriter, Australis was a reality and we took it to California in June 1967, hopeful of a launch into orbit within a few months.

In fact, it took two further years before Thor/Delta Number 76 soared off the launch pad at Vandenberg Air Force Base, California, on 23 January 1970, into a foggy morning sky. The Delta's main satellite was TIROS_M, or ITOS-1 when it reached orbit, an advanced polar orbiting weather satellite that NASA was launching for the National Oceanic and Atmospheric Administration (NOAA) – effectively the US Weather Bureau. I am sure that the scientists and engineers at NASA and NOAA anxiously monitored the progress of the launch, but not as anxiously as the little group that had built Australis, who were huddled in what the local media called "their tiny rooftop garret", the little room in the attic of the Physics Building that housed, apart from our tracking and recording equipment, a big, noisy electric contraption that pumped air (not air conditioning) into the windowless main Physics Lecture Theatre below.

We had reason to be anxious – the Delta rocket had a patchy success record: of the four previous launches, two had failed – and there was no Australis 2. We had no backup in case we ended at the bottom of the Eastern Pacific Ocean. This was our one and only shot at getting our satellite into space and realising what we had lived and dreamed about for five years.

The predecessor to Telstra, the PMG, had kindly provided a telephone connection between our "tiny rooftop garret" and the AMSAT-OSCAR people at Vandenberg, so we had a blow-by-blow description of the launch. The solid rocket boosters that circled the first stage of the Delta burnt out on schedule and dropped away. The first stage Thor ballistic missile shut down as planned, the second stage separated and ignited. So far so good – but that was what the man who fell out of the 15th story window said as he passed the second story on the way down – plenty of things could still go wrong. But Delta 76 was on its way and it kept going up. The fairing which protected the satellites until they were out of the atmosphere, came off as planned. The second stage continued to burn steadily. "ITOS separation!" we heard over the phone line. "The orbit looks good". So the weather satellite had been safely delivered into space. Now we had to wait. ITOS needed to be far enough away from the burnt-out second stage so that there was no danger of it colliding with NOAA's brand new satellite, and Australis could not be pushed away from the rocket with its two Henderson Federal Spring Works separation springs until that distance had been achieved. We waited – anxiously is too mild a word – for what seemed like hours, but was actually only a few minutes, for news from the AMSAT-OSCAR team at the launch site. It was so quiet in our control room atop the Physics Building (we had made sure that the "air conditioner" electric motor was switched off) that it seemed we had all stopped breathing ... perhaps we had! Then, the beautiful words we had waited so long to hear, "The NASA tracking station in Lima, Peru, has just received telemetry from the Delta second stage – Australis has separated and is in a stable, circular 1,400 kilometre polar orbit".

Five years of pent up emotions erupted in that "tiny rooftop garret" – there were shouts of joy, back slapping, hugs and a few tears. We had done it – a bunch of scruffy undergraduate students had designed and built a satellite from scratch – it was safely in orbit and it worked.

I could not have imagined, as I watched that faint star, Sputnik, glide across the night sky in October 1957, that a little over 12 years later I would look up at that same sky and know that I was a proud member of the team that had made our own satellite.

* * *

Owen wrote an Interim Report of the flight of AO5 and here it is reproduced in full, with his typing and spelling errors included, and in typewriter typeface:

AUSTRALIS OSCAR 5 INTERIM REPORT. by Owen Mace.

Australis OSCAR 5 is now silent, its batteries discharged after a working life of six weeks. The work of collecting and processing the thousands of reports from amateurs around the world begin. Before the last report on AO5 is written, many hours of computer time will have been used in the processing the data received from amateurs. Concurrently with this work, the next Australis satellite is being planned and designed.

AO5 was launched at 1131 GMT on January 23rd. in what could only be described as a fl awless, text book launch. One hour later, Australis separated from the Delta second stage and its two transmitters switched on. 5R8AS reported hearing the VHF beacon a few minutes later as the satellite came into range of his Malagasy Republic QTH. Minutes later as AO5 passed over Europe, DJ4ZCA and DL3OJ heard the 10 m beacon. In the following few orbits Australis was heard by amateurs in the UK, US, New Zealand, and, of course, Australia.

The response of US amateurs, especially, is staggering. Many thousands of reports have been received by the project to date. Some tracked every orbit in range throughout the life of AO5, some reported extraordinary antipodal propagation effects, and one even correlated the horizon sensor signals with cloud formations derived from weather satellite pictures, during the later part of the VHF transmitter's life. The patience of one brave soul is attested by this log entry;

> "On orbit 181/182, I could hear the 10 metre signal just about all the way around the world. I heard it for 95 of the 115 minute orbit, from very faint to fairly strong signal strengths."

WA2KS heard the 29 MHz transmitter commanded off during orbit 61 on January 28thÂ.

At the Project Australis headquarters station, VK3AVF[3], teams were organised to track the two high elevation passes each morning and afternoon. This vigil was maintained until the VHF beacon ceased transmissions during orbit 230 on Saturday 14th. February after 3 and a half weeks of highly successful operation.

The magnetic attitude stabilization system (MASS) worked very well also. The

---

3   Radio amateurs are assigned call signs with the first letters signifying the country of their origin: VK for Australia; W for the USA, etc.

satellite was soon locked to the earth's magnetic field by the MASS magnet. So accurate was this tracking that, by the 10th. February, the signal strength of the VHF beacon was appreciably lower as the satellite was south of the city, than when it was north. This was caused by the transmitting antenna becoming unfavourably directed by the earth's field as the satellite moved. Future designs will undoubtedly allow for this unplanned tracking accuracy!

The accompanying article by Jan King describes some of the preliminary results from Australis. In the ensuing months Project Australis will be analysing all the reports received in considerable detail to determine the affectiveness of the design procedures in order to incorporate modifications to the next satellite. Any reports are welcomed, so, if you have not already sent in your reception reports and station resume, please do so. The address is

> Project Australis(Telemetry)
> c/o Melbourne University Astronautical Society,
> Union House,
> University of Melbourne,
> Parkville, Victoria, 3052

* * *

The report went on to discuss plans for our next satellite. The remainder of my report is included in the next chapter.

Internal temperatures began at about 24C and rose steadily until it had stabilised between 47C and 49C by about orbit 78. These temperatures were somewhat higher than we had hoped for but well within the range that the electronics were able to cope with. The skin temperature, on the other hand, varied considerably as AO5 rotated in orbit from 45C to as much as 63C.

The spin rate, as estimated by the horizon sensors, was around 3.5 revolutions per minute during the first few orbits. After two weeks in orbit, the spin rate was so slow that it did not complete one revolution in a pass of twenty or so minutes! By this stage, AO5 was travelling with the VHF and command antennas roughly aligned north-south and in the direction of travel, as planned, so that fading of the VHF signal was reduced. The spin caused the HF antenna to rotate slowly and so there were pronounced 'nulls' in the HF signal strength.

We were surprised at comments reporting that the horizon sensors recorded clouds, snowy mountains and registered transitions of the sun and even the moon!

What I failed to mention was that there were failures, or at least some difficulties. First, the telemetry modulation on the 10-metre HF beacon 'dropped off sharply' at the end of the first orbit, making it very difficult to decode the data from the HF transmissions. (It was also reported that the failure occurred on the third orbit. No matter, modulation was much reduced for much of the flight.) It did appear about a month into the flight. The cause is unknown but my guess is a soldered joint somewhere had been shaken loose and a temperature change reconnected it later.

The Command System required considerably more radio power to activate it than had been calculated. Nevertheless, the HF transmitter was successfully commanded several times amidst much trepidation that we might not be able to turn it on again.

Plate 10.4 *A US Government Interagency Motor*
Pool car 'For Official Use Only' delivering Australis OSCAR A to NASA Goddard for testing.
Plate 10.5 *Australis OSCAR A about to be tested on a vibration table to ensure that it could survive the launch. Photograph courtesy of NASA.*

I was a little disappointed with AO5's operational life as I had hoped for a two- to three-month useful life, rather than the approximately 580 orbits over six weeks. This was largely due to the difficulties with the Command System and thus the 'high power' HF transmitter being left on for longer periods than we had planned.

Appendix B reproduces Jan King's Interim Report, which contains important information about the flight, as well as references to other articles and papers.

* * *

We might mention that we have pictures showing Australis was mounted very close to the second stage rocket motor. Perhaps this was to ensure that the NOAA people were satisfied that this non-

10.6

10.7

Plate 10.6 *Australis OSCAR A in its fighting uniform, the thermal control surface treatment. Note the locating pins, spring wells and tape antennas.*
Plate 10.7 *AO5 mounted on its 'baseplate' with antennas wrapped around it and mounted in the launch rocket.*

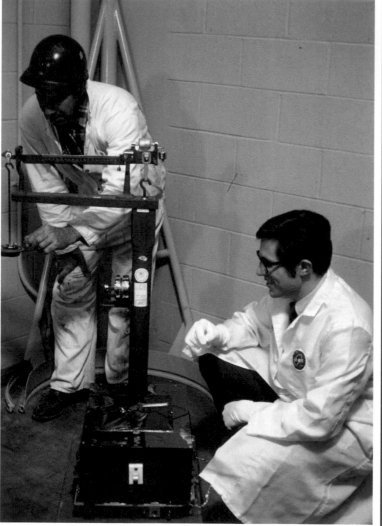

Plate 10.8

professional, foreign satellite was well away from their precious satellite. In any event, this would not have been a comfortable place to be, less than a metre from the shaking and thundering of the rocket engine.

Details of the Thor/Delta rocket, TIROS and the orbit are given in Appendix B.

* * *

It is worth imagining how things would have turned out if it had been Apollo 12 that had the in-flight explosion in November 1969 before the launch of Australis and not Apollo 13 after the Australis flight. The senior men in charge of NASA were the very same ones that approved the Australis launch and soon after dealt with the Apollo disaster.

Plate 10.8 *The pre-flight, weigh in. Jean Shrimpton, please note the white gloves!*
OPPOSITE PAGE
Plate 10.9 *Launch! The Thor/Delta takes 76 off into the mist from Vandenberg Air Force Base on 23 January 1970 with Australis OSCAR 5 on board. The University of Melbourne Gazette, Vol. 26, Number 2, 15 April, 1970. Reprinted with permission.*

UNIVERSITY OF MELBOURNE
GAZETTE

Published for University circulation, as occasion requires, by the Registrar for the Council of the University of Melbourne

Vol. XXVI, No. 2   Melbourne, 15 April 1970   *Registered at the General Post Office, Melbourne, for transmission by post as a periodical.*

Student Satellite in Orbit

Thor-delta 76 blasts off   Photograph, *NASA*

On 23 January 1970 the seventy-sixth Thor-Delta rocket was fired from the Vandenburg launch pad in California. The primary spacecraft was a NASA weather satellite, TIROS-M. However, to a group of students at Melbourne University the secondary satellite was of much more interest. This was AUSTRALIS OSCAR-V, a satellite designed and built by members of the Uni- versity Astronautical Society. As AUSTRALIS OSCAR- V separated from the Delta second stage, two radio transmitters were turned on, and their antennae un- folded. About one thousand miles beneath the satellite, an amateur radio station in Malagasy, reported that the very high frequency (VHF) transmitter was operat- ing. A few minutes later a weak signal from the high

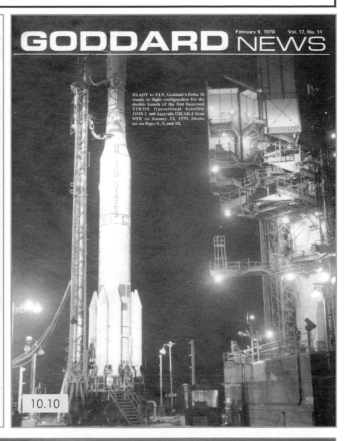

GODDARD NEWS
February 9, 1970   Vol. 17, No. 11

READY to FLY, Goddard's Delta 76 stands in flight configuration for the double launch of the first Improved TIROS Operational Satellite TIOS-1 and Australis OSCAR-5 from WTR on January 23, 1970. Stories are on Pages 8, 9, and 10.

Plate 10.10 *Ready to launch. Front page of the Goddard News magazine dated 9 February 1970. Credit: NASA.*

Plate 10.11 *Celebrations a few hours after the launch. Note the time on the wall clock.*

Plate 11.1 *Technology had moved on for Peter's radio teletype (RTTY) telemetry circuit boards for OSCAR 7 – CMOS logic chips – still working 40 years after launch.*

CHAPTER 11

Our Lives in Space

As noted in the previous chapter, Owen wrote an Interim Report on AO5's flight. The second half of the report described what we were thinking following the last transmission of our satellite. There is no date on the report, unfortunately. Once again, a typewriter typeface is used and all spelling and grammar errors are included to evoke the memories of those days:

Work is proceding with the design and testing of the next Australis. It is envisaged that there will be seven main areas of work; subsystems;

The main experiment, the repeater. While other groups, notably DJ4ZCA, are strongly in favour of linear translators, it is felt by the Australis group that the next step ought to be a hard-limiting FM system. Thus Project Australis is working towards a multichannel, channelised FM repeater. The plan is to use one receiver to mix to some convenient IF stage, split to several separated IF filters and detect down to baseband. Then each demodulated signals can then be used to frequency modulate its own carrier which is amplified and radiated. This system allows a number of advantages in that signal processing of the baseband is possible (eg speech compressing)and ALC) as well as removing doppler shift from the upgoing signal. It is presently planned to use about six channels receiving on 2m and transmitting on 432 MHz.

Telemetry System. It is hoped that a 60 channel telemetry system will be accomodated on the next satellite. Its output will be in the form of teletype signals impressed on one of the repeater transmit channels and operated on command.

Command System. A 35 channel command system will be incorporated to allow switching of receivers and transmit channels. This will allow great flexibility and will allow failed subsystems to be removed from the repeater system.

MASS. It is possible that a magnetic system similar to that carried by AO5 will be incorporated, although a gravity gradient stabilization has been mooted by Amsat.

Power Supply. A 6 watt solar powered battery is under investigation by Amsat, who will be responsible for the power supply.

Package. This is the responsibility of Amsat.

The design is for a lifetime of one year, to 85% confidence. It is anticipated that a prototype of the system will be flown on a balloon from Mildura in the near future, and interested amateurs are asked to listen to their broadcasts for further details.

Clearly we had a much more sophisticated satellite in our minds and Peter takes up the story of Australian satellite building below. However, Owen's family and his PhD studies drew him away from MUAS. He was sent on his first overseas field trip in 1971 to the National Science Balloon Facility in Texas. Imagine a three-month overseas trip these days without family, including a one year old infant! This was not the first long field trip, nor the last overseas field trip.

However, he was lucky enough to be seconded with wife Delia and son Nicholas to Prof Glenn Frye of the Physics Department at Case Western Reserve University for his postdoctoral year in 1973. What a year it was for them all. At Case they were building a totally new gamma ray telescope,[1] which was much, much bigger than the previous one. Glenn and Ruth Frye, their friends and colleagues were extraordinarily hospitable and it was a fabulous year for them all.

A memorable moment came for Owen when they were discussing a power supply for the new instrument. Batteries were considered but, being fascinated by the space shuttle that was being contemplated then, Owen knew that fuel cells were being studied. Glenn told Owen to make a cross-country telephone to California where the design studies were being conducted. Owen spoke to NASA's top expert on fuel cells who spent an hour with him discussing Case's requirement and the technology! Together they decided that batteries were the best solution and Owen discovered later that the Shuttle fuel cells cost many, many millions and so were utterly out of the question for his project. Wise advice from the NASA expert!

* * *

There have been a surprising number of space-related projects that Owen has had some connection with. Rather than describing each in detail, he has selected just two.

The mid-to-late-1980s saw an Australian Space Office formed to bring Australia into the space age. Among its many activities was to arrange for a group of Australians with experience and influence in space matters to visit the Russian Space Institute in Moscow and see some of their projects. The entire team was warned by "defence security" to be on their guard against being blackmailed for an indiscretion. The irony was that President Regan was visiting Premier Gorbachev at the same time as the visit and, by and large, the Russian security services had many other much more important matters to concern themselves with.

---

1   A relatively cheap but antiquated and dirty means of producing multiple copies of a document from a master sheet. that had been cut by a typewriter. The typist had to hit the keys heavily to ensure that the master was cut through so that the printing ink could flow through it to the paper. Considering the technology it produced a remarkably good quality product.

The group were warned that, should they return to their room after dinner to find an attractive naked woman in bed, leave immediately and report the matter to the Russian-speaking keeper of the room keys. As the group got to know each other during the visit, they began to ask each other over their caviar breakfast (literally) if anyone had met "Natasha" the previous night. There was general agreement, if not disappointment, that the answer was no.

The warnings were not without reason, however. One evening Owen was sitting in a lounge and pulled up his wing back chair to hear a conversation more clearly. It took a bit of force to move the chair … and to break the microphone wire!

Nevertheless, it was a most interesting and indeed enjoyable ten days in Moscow.

The Space Office continued for over a decade and among the contracts that it funded was a project to build test equipment for the Advanced Along Track Scanning Radiometer (AATSR) that flew on the European Envisat satellite (2002 to 2012). This instrument, which had been co-invented by another of Vic Hopper's PhD students, was the latest in a series of similar instruments to measure sea surface temperatures from space. The significance of the measurement is that sea surface temperatures can be measured from space to an accuracy of around 0.3C over a much wider area than is possible by any other method and so represents a figure for the average temperature of the Earth itself. Furthermore, the multi-decade series of measurements are highly significant for climate research.

Much to the chagrin of the "traditional" winners of such space projects, the company that Owen was working for, Aspect Computing, won the contract for the AATSR Ground Test equipment. To further add to the annoyance of Aspect's competitors, they did so profitably and on time. The state manager of Aspect encouraged numbers of successful bids for projects that would not ordinarily have been the province of an information technology company and yet were largely software production, which was their strength.

Owen says he has been extraordinarily lucky in his professional life to have seen and done all that he has, including a trans-Atlantic Concorde flight as part of a business trip. Let us now read how others in our group have lived their lives.

* * *

Richard Tonkin

At the time of the launch, Richard had been married to his wife Pauline for just over a year. They were living in a flat in East Malvern, she was pursuing her musical career with the Melbourne Symphony Orchestra and he was working full time at the Weather Bureau in Melbourne, in charge of receiving and processing pictures from American weather satellites. Frankly, he was over Australis, Pauline was pregnant with their first child, Karen, by October 1970 and they had bought a block of land in Ringwood, in suburban Melbourne, and started to build a house.

So, he did not follow Peter's exploits with OSCARS 6 and 7 – perhaps if some of the Australis team had stayed together, we could have gone on to create a business building small satellites, negotiating launches and selling space on our satellites to the universities and the CSIRO for experiments. But it was not to be.

In 1972, Richard finally finished his Law course, qualified as a lawyer and got a job with a firm in the northern suburbs of Melbourne, where he recently retired as the senior partner.

But space was still in his DNA. The family moved from Ringwood to acreage in Smiths Gully, north of Melbourne, had two more children, Richard worked hard in the law firm, as did Pauline in her musical career. His interest in space was renewed when he joined a tour led by an American group to Russia in 1992 to watch a manned Soyuz launch to the Russian space station Mir.

This was just after the Soviet Union had collapsed, Russia was opening up to the world and anxious to gain foreign currency. It was an amazing trip – the group flew from Moscow to Leninsk, the service town for the Cosmodrome at Baikonur, and then by bus to the spaceport. This was nirvana for space enthusiasts: the location of Baikonur had been a closely guarded secret throughout the Cold War, yet there they were, a group of Americans, two Canadians, two Belgians and Richard, welcomed into the very heart of the now Russian space programme.

They watched the Soyuz launch from 500 metres away, only an old post and wire fence separating the viewers, who included cosmonauts and generals, from the rocket. Being that close, the vibration from the launch seemed to go right through Richard's body – he was sure he could hear his teeth rattling.

There was a party that night to celebrate the launch, where the vodka flowed freely. Richard vaguely remember standing on a table with his arms around the two Canadians, singing their national anthem, O Canada. He swore off vodka that night and hasn't touched a drop since.

The next day, the group went to the launch pad of an unmanned Proton rocket which was to be sent into orbit the next day with a Russian communications satellite. They were allowed to climb the ladder which spiralled up the side of the rocket to inspect it. Imagine asking to do that at Cape Canaveral!

They then had permission to go to the pad from where the Soyuz had been launched the previous day. As they walked around, picking up pieces of burnt concrete as souvenirs, their guide showed them marks painted in the Cyrillic alphabet on the side of the launch gantry – the first one read "Sputnik 1" and further down "Vostok 1". This was the very launch pad from which the first satellite and the first man in space were sent into orbit – truly hallowed ground for space enthusiasts.

The three days at Baikonur ended with a visit, first, to a huge hangar where the second Russian space shuttle, Buran, was being built – the programme was cancelled because of the cost to the struggling Russian economy. Then to the little house where Yuri Gagarin spent his last night before his historic flight in 1961, with the Chief Designer for the Soviet space programme, Sergei Korolov.

The tour ended back in Moscow where, surrounded by more cosmonauts and generals, they watched on giant TV screens, the docking of the Soyuz with the Mir space station. It was a remarkable trip which whetted Richard's appetite to visit other launch sites around the world.

In late 1992, he got himself invited to the launch of an Australian Optus communications satellite from Xichang in China, In 1993 Richard took some of the Americans and Canadians from the Russian trip to French Guiana, on the east coast of South America, for a European Ariane flight. In 1997, he was invited to a Japanese H-2 mission from the beautiful island of Tanegashima, carrying a weather satellite and scientific payload. 1994 saw him at Cape Canaveral for a Space Shuttle flight and he returned in 2006 for the launch of NASA's New Horizons spacecraft to Pluto.

* * *

Fedsat was announced in 1996, to celebrate the 100th anniversary of Australia becoming an independent nation. There was no Fedsat 2 and no space agency. Now, over twenty years later, there is a further push for Australia to become a space-faring nation. In September 2017, at the International Astronautical Congress in Adelaide, the Australian government announced the creation of an Australian Space Agency (the ASA). With a staff of just twenty and a very limited budget, it remains to be seen what the ASA can actually do. Hopefully, it will be properly supported by both government and industry so that Australia can at last, 50 years after Australis, truly become a spacefaring nation.

Whatever happens, Richard will continue to be fascinated by space research and development and by the enormous potential that it offers to Australia and to the world.

PETER HAMMER

After Australis OSCAR 5 was launched, Peter was still very interested in amateur satellites, but the rest of the AO5 crew went off to do other things. With the support of AMSAT and the US Embassy in Canberra he was receiving thick weekly packages of information and components which probably didn't go down at all well with Prof. Hopper. Peter went on to develop the command system for OSCAR 6. This was built using the very new first generation of integrated logic developed by Fairchild. It was simple RTL (resistor-transistor logic) and packaged in epoxy sealed ceramic disks. Not qualified for space use; however, it all worked perfectly. From here, the next project Peter got involved with was the Radio Teletype (RTTY) telemetry system for OSCAR 7. This was built using military grade CMOS logic[2] made by the RCA Corporation. Again, a brand new technology which his lecturer in electrical engineering said would never replace transistors! He was totally wrong as it is the standard system used nowadays in every integrated circuit and image sensor. These chips were extremely sensitive to static electricity, which meant that special precautions had to be taken to make sure everything, including me, was properly earthed when working with these chips.

Many experts predicted that these circuits were totally unsuitable to space use and would be lucky if they lasted a few weeks in the harsh environment of space. Again, they were wrong as the RTTY telemetry system is still working after around 40 years in orbit and indeed the OSCAR 7 spacecraft holds the world record for the longest working satellite in Earth orbit. After the launch of OSCAR 7 Peter was pretty

---

2   Complimentary Metal Oxide Semiconductor logic, now the mainstay of digital electronics.

burned out and lost interest in amateur satellites and amateur radio and got involved with life and to a degree with his other hobby, photography. The boards for OSCAR 7 were made commercially through a company that was prepared to make them for him for free. All components had to be hand-soldered into place.

After finally finishing his PhD at the end of 1972 Peter started work at LaTrobe University where he was helping postgraduate students in the Physics Department.

For OSCARs 6 and 7, Peter was working at LaTrobe University and had a very good 8"x10" camera in the electronics workshop. Technology had now moved to laying out the track pattern at twice full size using black precision cut tapes which came in a range of widths. The tape was placed on a transparent plastic film, which was then photographed and reduced in scale to the correct size. This produced negatives on large scale film. Copper clad board was coated with a photosensitive resist (or you could purchase pre-coated board) in a darkroom lit with red light. The copper clad board was then sandwiched between the top and bottom layers of the track pattern and exposed to white light to burn the track pattern into the resist. The exposed boards were then etched in ferric chloride to dissolve the unwanted copper and the required holes were then drilled.

Peter married, had a son and eventually divorced. He has travelled considerably over the years and has now visited about forty countries and all seven continents. After a trip to Africa in 2005 Peter joined a camera club and started entering competitions, the Australian Nationals to begin and later, international salons. Since then he has accumulated around 2,200 acceptances in international salons, 137 awards in international salons and a total of 17 Gold Medals. He is a Grand Master of the Australian Photographic Society.

(Note from the authors: It is well worth a visit to his website, www.peterh.photography.)

STEVE HOWARD

After completing an electrical engineering degree, Steve joined the research laboratories of the Post Master General's Department, working first on microwave radio propagation in northern Australia, then on the design, construction and installation of solar radiometers to measure the attenuation of microwave radiation at 11 and 15 GHz in tropical conditions. Data from this work was used in the design of satellite communication systems.

In 1981, Stephen started his own company, Tain Electronics Pty Ltd[3] where he has developed a wide range of monitoring and control equipment for use in education, environmental research, agriculture and industrial applications. His latest product is a range of data loggers for monitoring soil moisture in the agricultural industry.

---

3   www.tain.com.au

Rod Tucker

Rod completed his PhD in Microwave Engineering at the University of Melbourne in 1975 and then continued his research at the University of California in Berkeley, Cornell University, and the Plessey Company in the UK. In the late 1970s, he became fascinated by the emerging field of optical fibre communications, and joined Bell Labs in New Jersey to pursue a career in this area. He was located at Bell Laboratories, Crawford Hill, the site of some of the very early Telstar satellite experiments. In the early 1990s Rod came full circle, and returned to the University of Melbourne as a Professor of Electrical Engineering and Head of Department where he worked until his retirement in 2014. In that time, he taught many courses and supervised approximately 40 PhD students. One of the highlights of his time at the University of Melbourne was his participation in a Panel of Experts that advised the Australian Government on the establishment of the National Broadband Network.

Jan King

We've read that Jan King was a young engineer at NASA Goddard where he was participating in testing of all manner of NASA spacecraft, earth orbiting, lunar landing and interplanetary. He met many technical experts and, because of his involvement with AMSAT and Australis, he also met many of the most senior managers within NASA. These meetings most definitely helped his career. As he says, it helped him "see a bigger picture of the world … it quickly formed my career".

The AMSAT organisation advanced "tremendously" from its decision to fly Australis. AMSAT has been involved with over one hundred satellites and not just amateur radio satellites, but satellites built by universities and others. It has flown on rockets from every launching country, including the former foes of the US: the Soviet Union, now Russia.

Jan now lives in Australia and is still mesmerised by the excitement of space and spacecraft; he truly is an engineer of the first order. He continues to work on space projects and, as Owen writes this, is in Svaalberg, Norway where he is establishing a highly specialised receiving station for a series of satellites he is working on.

Reflecting about Australis and his life in Australia, Jan is less than complimentary about the risk culture in this country and the fact that we still have no space programme of any kind, despite many attempts to interest various Australian governments. He refers to Canberra as 'Can't berra'.

In an interview, Jan likens the engineering of spacecraft to the Earth itself. The creative effort to design a spacecraft to measure, or do something totally that has never been done before, has parallels to the challenges facing all the peoples of the spacecraft we call Earth. New ideas are needed to engineer new solutions to new problems on a new spacecraft and on spacecraft Earth.

For the first edition Owen wrote:

> As we are finding with our young Australian engineers and people, they are no longer waiting, as we did after AO5, for governments to take a lead but are just getting on with developing their ideas, hindered unfortunately by red tape. Get out of the way, "Can't berra", and let these people flourish.

We are seeing this happen with school children around the world demanding real collective action on global warming from our so-called leaders. As Australian would say, good on yer, Greta Thunberg.[4]

* * *

OWEN'S VW

The "Vee-dub" served Australis and Owen's early married life well. However, loading an infant into the back seat of a two-door car is difficult and so he traded it in for a larger, four door car – still German but this time second-hand.

Plate 11.2

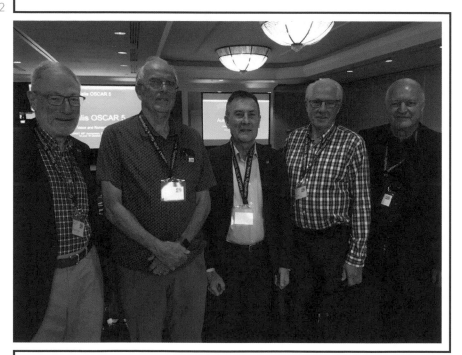

Plate 11.2 *Members of the Australis, OSCAR and AMSAT teams at the AMSAT 50th Anniversary Symposium, 19 October 2019. From the left, Richard Tonkin, Lance Ginner, Owen Mace & Jan King.*

---

4   https://en.wikipedia.org/wiki/Greta_Thunberg accessed October 2019.

A Vehicle Built For Two by Robert Goss

Delta Project Office
NASA Goddard Space Flight Center
February 1970

Sung to the tune of "A Bicycle Built for Two"
Key: various.

1st Verse
This is NOAA Owner of the primary payload,
TIROS M, refusing to allow AO5 to ride on the
Delta rocket with TIROS M.

OS-CAR, OS-CAR give me your an-swer, do.
I'm un-will-ing to-share my ride with you.
You'd probably block my vision,
Or cause a space collision.
So, I'll be damned if I'll be crammed
On a vehicle built for two.

2nd Verse
AMSAT's response to TIROS' refusal:

TI-ROS, TI-ROS, here's my reply to you:
I'm approved by N-A-S-A HQ.
So, don't be unkind or bossy,
With this hitch-hikin' Aussie,
And I'll look neat in the rumble seat
Of a vehicle built for two.

3rd Verse
NOAA lamenting NASA's decision at fly Australis.
The third line refers to the Delta Office wanting
to prove Delta's improved lift capability with six
strap-on solid rockets. This meant that there was
more than enough fuel to carry the extra (tiny)
mass of AO5 into orbit.

Del-ta, Del-ta here's all I have to say:
If he bumps me there will be hell to pay.
Six solids we could not pre-vent
This ham adds to our tor-ment.
We never thought that we'd get caught
On a vehicle built for two.

4th Verse
AMSAT encouraging NOAA that all will be well.

TI-ROS, TI-ROS, not a thing can go wrong,
Be a hero - let OSCAR tag along.
And be a content precursor,
It could have been much worser -
We could, you see, have carried three
On this vehicle built for two.

5th Verse added by Pauline and Richard Tonkin
for the 50th anniversary

TI_ROS, OSCAR see what I did for you
In an orbit almost too good to be true
You're here for ten thousand years,
There was no need for tears
So, you see, Mr. Douglas made me
A vehicle built for two.

CHAPTER 12

The Melbourne Space Project Half a Century Later - A Students' Journey to Space

Authors' Note: This chapter was written by Andrew Wetherell, Steven Despotellis, Anna Bohacova, Gabriel Abrahams, Rowan Skewes, Benjamin Ding and Tim Allison-Walker for the Melbourne Space Programme.

INTRODUCTION TO MELBOURNE SPACE PROGRAMME

In 2014 a group of students banded together with a mission, to design and launch their own satellite in Australia, the first in 50 years since OSCAR-5 in 1970. The team combined an engineering professor with a mix of enthusiastic engineering, law and business students, excited by the opportunity and challenge ahead. This was the founding moment of the Melbourne Space Programme (MSP), an organisation dedicated to launching the next generation of technology pioneers. MSP continues to run student-led satellites and space projects as a training vehicle to provide professional experience to students, preparing them to make an impact in the local space and technology sectors.

ACRUX-1, the first satellite mission, was a critical component in preparing MSP to undertake this journey. It laid the foundations and served as a proof of capability, demonstrating student's aptitude for building a satellite in Australia and developing the processes internally to enable more satellite projects, re-igniting the passion of students for the space industry in Australia.

Plate 12.1

Plate 12.1 *The Melbourne Space Program Team*

The ACRUX-1 mission was a proof of concept and capability for MSP, demonstrating Australian students' ability to lead the design and launch of a satellite, however it didn't start that way. In 2014 when the team were first formed there was a different goal, to launch a satellite with an inflatable antenna to run high-gain communications on a nano-satellite.

The primary focus was on designing a novel antenna with a thin material that could be folded and securely packed into a cavity in the satellite during the launch phase. It was imagined that once in orbit, a special powder would sublimate into a gaseous state, forcing the material out of the satellite and blowing it up like a balloon. The concept had been demonstrated previously on larger satellites. However, such a feat would have been a world-first on a nano-satellite. This is where MSP hit the first set of challenges.

The technical difficulties involved to achieve the design of an inflatable antenna were significant. Not only was this an extremely complex and novel payload, but the team comprised mostly students with little or no understanding of space technology at the time and few local space experts to draw support or advice from. The cost of such a mission was also unsupportable, with a focus on the payload and mission success there was little funding opportunities for the team at the time and no government support for space endeavours with the founding of the Australian Space Agency happening only a few years afterwards. The lack of feasibility and overflow of ambition was recognised and during a few months of deliberation the team concluded that the project needed to be rescoped and focus on first delivering a proof of concept, a demonstration that students in Australia can design and launch a satellite.

Plate 12.2 *Early development of satellite boards*

Plate 12.2

The re-scoping of the satellite to a 1U (10 x 10 x 10 cm) proof of concept with the goal of communicating was a critical phase for MSP and the mission. This came with the spin-out of MSP into its own organisation and shifted the focus of the organisation from launching a satellite to using space technology projects as an incredible training vehicle for students – delivering exciting projects and training simultaneously. This was also when the first satellite earned its name, ACRUX-1, a tribute to the brightest star in the Southern Cross, a reminder of the ambition MSP intended to develop as well as the project being a crux point for the organisation, the turning point and re-focus of what could be achieved together in the organisation.

Plate 12.3

THE ACRUX-1 MISSION

The goal for the ACRUX-1 mission was to launch a 1U satellite designed by students with the stretch goal of being able to communicate with thesatellite. There were significant challenges and learning which needed to happen in order to achieve this mission, including learning how to design all avionics on-board, how to design a chassis and integrate a satellite which can meet the stringent qualification testing requirements and navigating a complex regulatory environment. All of this was to be done by students with industry volunteers where possible, guiding best practice and advising on challenges as they arose, combined with the need to continue to run the organisation as a whole and establish a sustainable company which can continue to train students while being run by students themselves in university at the time.

TECHNICAL BREAKDOWN

The approach to developing the satellite was to breakdown the different sub-systems involved and have students lead each team with leaders coming together to do systems engineering and general planning. We initially ran on large waterfall project management approach with complex change process, requirements gathering and management of tasks and team integration. The different subsystems which were required to be developed were Flight Systems, Flight Vehicle, Telecommunications, Mission Operations, Software and Power.

FLIGHT SYSTEMS

The flight systems subsystem was focused on building the capability to control the spin rate and orientation of the spacecraft relative to the Earth. Originally the goal was to use reaction wheels for this but for space consideration within the satellite this had to be re-scoped to developing our own magnetorquers as an active attitude control system and hysteresis rods for passive control. While the ability to control the rotation of the satellite was not critical to mission success, we saw this as a crucial capability building requirement which would set the foundations for future missions in carrying out more complex objectives where control of the spacecraft orientation is important. Included in the

Plate 12.3 *The evolution of the satellite*

subsystem teams' responsibilities was to lay out all the sensors on the satellite to measure orientation as well as a general health suite which would provide data on how different parts of the satellite are functioning once in space, and writing all code for the subsystem.

As a result of time and funding constraints, the team had to be creative in testing the system when facing challenges such as an inability to reproduce free-fall rotation on Earth. A major strength that they were able to leverage today was the inclusion of advanced simulations to test systems which have previously been unavailable to predecessors such as the Australis OSCAR-5 team. Intensive simulations using the onboard computer to control simulated rotation in real-time was implemented as a method to test the magnetorquers. In all our testing this showed the spacecraft was tumbling slowly on orbit which would later be tested against the live launch and operation of the satellite to improve future simulations. Collecting enough data to fully characterise the change in tumbling however was out of scope and so not something the team would be able to do entirely.

Plate 12.4

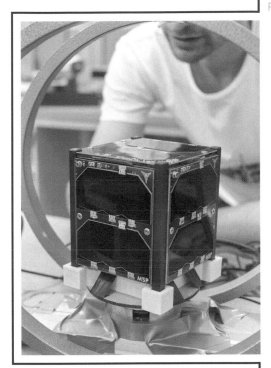

FLIGHT VEHICLE

The flight vehicle team had a slightly higher-level scope than other engineering teams and was primarily responsible for the satellite as a whole, including the development of the satellite chassis, electrical and mechanical integration, thermal control systems and the management of the satellite's technical regulatory compliance. The main in-depth design tasks revolved around the chassis production and thermal analysis, whereas the other responsibilities, such as establishing an electrical connection standard for the satellite and managing mass/volume budgets required a high level of inter-team communication and collaboration.

Plate 12.4 *Testing the satellite control system*

There were many challenges faced by the flight vehicle team which included managing the mass and volume budgets for the satellites. This was particularly difficult as most components were custom developed by MSP and constantly being tweaked or revised. The 1kg, 1U cubesat form factor does not lend itself to a particularly spacious satellite and it was only by maintaining a very accurate, up-to-date CAD model at all times that we were able to avoid nasty surprises coming in at final integration.

The aspect that really cramped the internals was the decision to use Molex board-to-board connectors as the electrical interfaces rather than the CubeSat standard PC104 type connectors. This was primarily due to reports of faulty PC104 connectors coming loose during launch, however the resulting wire harnesses required with the Molex connectors made integration a painstaking process. Through sheer determination (and many hours crimping wires) the team fitted everything together and know that in the future a better method can be found.

The final critical issue was to do with the thermal control system. Accurately modelling the heat transfers within the satellite and its interaction with space was difficult, with various results being identified depending on the assumptions used. Most simulations seemed to agree with observation CubeSat data that the internals sit at about room temperature once all electronics are up and running. This, combined with the very limited power budget and available space, resulted in the decision to forego a proper thermal control system and just rely on a simple battery self-heating resistor circuit. Unfortunately due to the power performance observed during final testing it was decided to remove even this circuit to help avoid brownouts on the satellite.

Overall the Flight Vehicle team managed to accomplish the vast majority of their objectives. The chassis was solid and successfully passed environmental testing and survived the launch, the satellite integration was relatively smooth (if somewhat fiddly) and ACRUX-1 was deemed compliant with all required CubeSat standards. The most important verification for the Flight Vehicle team however is that ACRUX-1 safely made it to space and operated well enough to successfully make contact with the ground.

Plate 12.5

Plate 12.5 *In-house testing of the satellite*

TELECOMMUNICATIONS

The objective for the telecommunications systems was initially to be a high data rate multichannel system designed entirely by the student team. However, as the team progressed with their design and went through an intermediate design review it was clear that the students most familiar with the design were graduating and the new telecommunications team were not confident enough yet to design a bespoke radio. A review concluded the best path forward was to build a low data rate off-the-shelf system to give the overall mission the highest chance of success, making the telecommunications the only avionics to be off-the-shelf.

The radio and antenna were selected on a number of criteria including flight heritage, complexity and cost. The next stage was to design the PCB which interfaced the radio and antenna with the rest of the satellite, and included health sensors (temperature, voltage, current) to monitor performance. Finally the board was integrated with the other subsystems, and lots of debugging performed.

In addition to assembling the satellite electronics, the team needed to design and verify the radio frequency system. Critically, a link budget was calculated which estimates the signal to noise ratio and hence the data rate achievable during a satellite pass. The team also tested every aspect of the system as much as possible, for this the amateur radio community were extremely helpful in mentoring and lending equipment.

MISSION OPERATIONS

The objective for the mission operations team was to be able to receive encoded transmissions from orbit, decode the data and display it in a legible format. Included as a stretch goal for the team was to be able to send various commands to the satellite to enable additional functionality. Because of timeline and resource constraints being able to decode a packet down from the satellite was deemed more important where we would start the satellite in limited functionality and only if working could commands be sent to then boot up further functionality for testing purposes.

The mission operations team were mainly focused on software, paying particular attention to the ability to encode and decode transmissions and implement Forward Error Correction (FEC) redundancy. This included the ability to interface with the telecommunications link and designing a display which would let the team monitor the satellite and visualise the data in a way which could be easily interpreted by the team for analysis. There were significant challenges in developing new software to solve particular issues related with the telecommunications interface. As part of the operations there were repeated iterations and alterations to scope, partially driven by project management style changes and a lack of depth on initial systems design and understanding of desired functionality. The implementation and testing of FEC on both ground and satellite posed significant challenges and then the verification of ground control systems.

One of the key approaches to overcoming these challenges as the development of a "flatsat" or breadboard, and engineering qualified models. The flatsat was one of our prototype satellites laid out on a table, board by board to allow for easy testing and interaction with the satellite and for running on-going tests and simulations. The engineering qualified model was our satellite which had passed initial testing requirements at a testing facility to ensure that we are ready for developing the flight model and going to meet all qualification requirements for launch and operation in space. The flatsat and engineering qualified model enabled more thorough testing of the telecommunications link and associated mission operations control which was critical. The team combined more with the Software team as well, providing much needed expertise in the satellite code and allowing the team to implement

FEC directly into the onboard software. Testing of the ground station control systems via tracking existing satellites prior to launch was key as well.

SOFTWARE

The software team were focused on the on-board flight software as well as the integration of all other teams code and general architecture for software. This was a significant responsibility, requiring the organisation of different subsystems code and integrating everyone's work into a consistent format and architecture that would enable all subsystems to function as intended.

The key responsibilities for the software team were to develop the ability to control the radio uplink/downlink, collect health data, perform system health checks, recover from potential system failure and implement the de-tumbling algorithm and orientation control. There were significant constraints on the team including limited RAM, interfacing a large number of sensors with different data formats with a central computer and the potential for an energy negative state on the satellite during operation. One of the hardest challenges was also the coordination piece, designing software in a way which is easy to test, dividing work between team members and having the opportunity to do the testing on hardware.

To overcome these the team followed good object-oriented principles and software design. Unit tests were introduced and the software was separated from hardware inputs. Again the flatsat and engineering qualified model were critical in testing using remote computers to test more actively. A measurable framework was important for establishing the health checks and a custom boot sequence in the scenario of energy negative state was critical to ensure that the satellite would follow an intelligent flow of functionality bootup to maximise the chance of communication. This led to successful beacons received from the satellite.

Plate 12.6

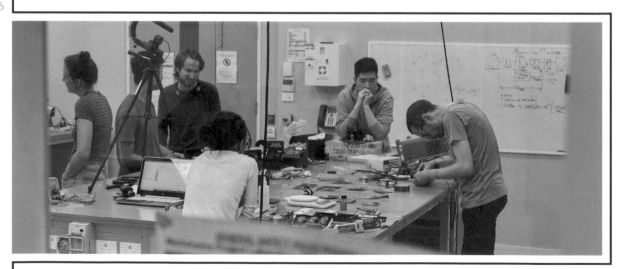

Plate 12.6 *Build day in the workshop*

PROJECT MANAGEMENT

A critical component to the successful development of the satellite as a whole was the project management approach. Initially we implemented a traditional approach using complex requirements gathering, a waterfall chart to manage schedule and visualise the interdependencies between teams as well as a formal change requirements process.

Very early on the team hit significant challenges with the project management approach which was driven from two key factors, the requirement of managing a 60 person volunteer engineering team and the immense amount of uncertainty in the entire design of the different subsystems of the satellite. With so much uncertainty in how to design a satellite and little local expertise to get support from attempting to gather all requirements and map out the satellite was too complex and inaccurate, coupled with a formal change process which was inhibiting the regular shift in priority of work and scope of the avionics. Understanding the interdependencies between teams was also extremely complex, focusing on all the requirements overlap with a large group of students.

There were two key events which helped to resolve these extremely large challenges which the satellite would not have been developed without overcoming. The first was the introduction of a systems engineer from industry who helped the team to refocus their attention to key functionality rather than all the requirements of the subsystems and set out a minimum viable product approach to reaching core functionality which helped to focus teams on key design and testing and defer commitment to stretch goals until these designs had been reached. This also helped to build a much better understanding across the team at a high level on how each subsystem integrates together when focusing on the functionality that had to be achieved rather than the traditional scope of all subsystems. The second event which was crucial was the shift in project management approach to the use of agile principles and the introduction of Scrum project methodology across the engineering disciplines. This focus put much more emphasis on communication and team dynamics, removing formal change processes and welcoming change as part of the learning experience.

The agile transformation saw a significant shift in the momentum of the team. Communication became a lot strong across the entire organisation, focus on tasks and working towards the minimum viable product was crucial and welcoming change as part of the process especially with so much inherit uncertainty in what the team were doing was vital if the team were to learn fast enough to deliver a satellite. There were a number of benefits from this shift and overall it contributed immensely to the successful launch of the satellite.

LEGAL

Understanding and complying with our legal obligations was one of the major challenges we faced with the ACRUX-1 mission. We first needed to understand what legal obligations we had. After approaching a number of law firms and individual lawyers for advice we found that few had any understanding of the legal process to launch a small satellite under Australian jurisdiction. There were a few companies who launched satellites under Australian laws, but all were large corporations or government organisations and the legal services providers we approached were sceptical of a small satellite project being successful in securing all the necessary permits. So started our legal journey. MSP had a small regulatory team at the time, made up of four law students and all of us jumped in to learn about this area of law.

The main legislation governing satellite launches is the Space Activities Act 1998 (Cth). At the time we started, the legislation has not gone through any major alterations since 1998 and was from our perspective ill-suited to the advances in satellite technology over the two decades.

FREQUENCY ALLOCATION

One of the crucial legal aspects of launching a satellite is acquiring a frequency to communicate with the satellite. There is a limited spectrum which can be used and without international cooperation, operators would face disruptions in communications. Therefore, the coordination and allocation of frequencies happens at an international level and is governed by the ITU (International Telecommunications Union). Individual states have their own agencies that then communicate with the ITU. In Australia the designated agency is the Australian Communications and Media Authority, ACMA. We contacted staff at ACMA for help with the application for a frequency. The entire process is broken down into several steps and it took us some time to understand the exact steps we had to take to successfully receive a frequency allocation.

There were lots of challenges involved in acquiring our ACMA license and major time delays which were difficult to predict. Because in Australia there are very few satellites launched and even fewer by small companies who go through ACMA there was a lot of uncertainty on both sides as to what steps had to be undertaken to successfully acquire a license to communicate. We worked with the International Amateur Radio Union and Wireless Institute of Australia to coordinate the acquisition of an amateur communication frequency license which required communicating with governments globally to ensure that our satellite would not interfere with their communications.

Plate 12.7

Plate 12.7 *Giving the satellite some artificial rays*

INSURANCE

One of the largest risks to the success of ACRUX-1 would be getting insurance. When a space object causes damage, it is the State in which the object is registered rather than the individual company that is liable for any such damage. If ACRUX-1 damaged another satellite from a different country, it would be Australian Government, not MSP that would be on the hook for any monetary compensation. For that reason, the Australian Government requires operators of satellites to acquire extensive insurance on their satellite and mandates that any damage would be paid out of the insurance rather than by the Government. The insurance cover needed was at the time of the ACRUX-1 mission set at $750 million.

For a student-led small outfit like MSP, it proved near impossible to arrange for an insurance cover for that amount. In Australia this was particularly difficult, because insurance pricing is based in risk and local insurance companies don't have experience with insuring spacecraft, and so no basis for pricing the risk associated with the satellite. Our only alternative with was to work with the government using an insurance waiver from the Space Activities Act which specifies that the insurance requirement can be waived by the Minister.

As we begun this process to apply for the insurance waiver we were met with another roadblock, the process shifted mid-way through application from the Minister's office to the newly founded Australian Space Agency. This delayed the application and after months of working with the Agency we managed to get the waiver days before we were expecting to transport the satellite based on the original schedule for launch.

LAUNCH CERTIFICATE

To launch a satellite under Australian jurisdiction, one must get a 'launch certificate' which is a permission from the Australian Government to launch a satellite and is mandated under the Space Activities Act. The certificate normally costs $100,000 but if an organisation is a declared scientific or educational organisation, it only costs $100. MSP didn't have the finances for the normal certificate application costs, so we applied to the Minister to be officially declared a scientific or educational organisation, thereby saving $99,900. We argued that we qualified as an educational organisation due to our mission and activities and were successful in securing the Minister's declaration. We also received the launch certificate with time to spare before our launch.

As our launch shifted from originally being in the U.S to New Zealand we also had to apply for an Overseas Payload Permit to be allowed to launch in New Zealand. Thankfully, the local space agency was very collaborative and processed the license quickly.

THE LAUNCH

An essential part of achieving mission success would rely on launch success and choosing the right provider who we can meet qualification requirements for. In Australia learning about the best launch providers and the process to successfully launching was not well known with little local expertise to guide what steps we could take.

After months of deliberation we had chosen to use a launch broker who acts as an intermediary between the launch provider and satellite operator. We locked in a launch for November 2017 on Falcon-9 rocket. This was a crucial part of the mission, setting a hard deadline on when we had to deliver the satellite by. At the time, the team believed that this was an achievable feat and so put a significant amount of budget towards the launch. Falcon-9 was chosen for its heritage with the PSLV and Rocket Lab only just beginning, to maximise the chances of success. However, during significant progress into the satellite development we were given news that the primary payload on the Falcon-9 had shifted the orbit to one in which we would no longer de-orbit within the 25 year international regulation requirement. We were given 1 month to re-negotiate a new launch contract during the teams exam time and the decision would force a re-design of elements of every subsystem on-board the satellite.

During frantic planning sessions we re-negotiated the launch to a Rocket Lab Electron rocket which would require re-designing the power system and subsequent functionality across the satellite to deal with shifting from a sun synchronous orbit to a polar orbit which would have the satellite in eclipse for half of the day. This design feature ultimately would take too much time as well to change within the 3 months provided to get to the new launch including all qualification testing and so a new launch was chosen in late 2018 to provide time for the redesign – something which ended up continuing to be a challenge with a power negative state at launch.

With the new launch date our target was to deliver the satellite by November 2018. The team worked around the clock and managed to bring everything together through three iterations of the satellite.

After 5 years of work, lots of failure and a large cohort of students rolling through the programme we were ready to launch. It was December 2018, the satellite had gone through its finishing touches and we had met all qualification testing at the Advanced Instrumentation and Technology Centre. It was the final days leading up to transport and we received the rest of our legal licenses including the export license and confirmation of our communication license, the satellite was ready and then … launch delay. The launch schedule had slipped, so transport would not happen until 2019.

The window came around again to launch and another delay. In the end, after several delays, launch shifted to June 2019 – the time where we would get to launch. To transport the satellite over to New Zealand we had to first take it over to the U.S where it would be integrated into 3U pods with other teams satellites, before being transported

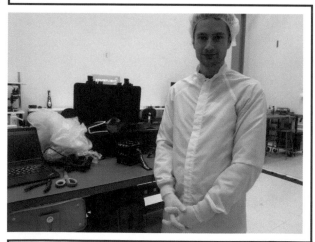

Plate 12.8 *Finally got to the States*

Plate 12.8

over to New Zealand. To get it there was as simple as one of the team members carrying it as hand luggage on the plane – with all the appropriate approval of course. The transport and integration were seamless and we were all ready for launch, finishing up work on the ground station to prepare for communication with the satellite.

Plate 12.9

Plate 12.9 *Electron launch capture by team in NZ*

The day came around, we were ready for the launch, and the launch was postponed. The next day rolled around, again all prepared for the launch, and it was postponed. We thought the third time's the charm, but again, postponed. On the fourth day and coming to end of the launch window all systems looked a go. We all waited remotely operating our ground station in preparation for successful launch, and the Electron launched.

Success, everything had gone as intended with the launch and our satellite had been deployed from the pod now waiting for communication. The first pass happened over Melbourne, and nothing. The next pass happened, and still nothing. Then the third pass and still nothing. Until, an amateur community of ground station operations, SatNogs, reached out through Twitter to let us know that they had picked up ACRUX-1 communication and we received our first packets. Success! It was an emotional moment for the whole team and thanks to an incredible community of global ground stations operated by enthusiasts that we had communication.

The satellite was operating beyond expected capability, not only sending packets but all subsystems appeared to be functioning as intended – something we thought to be improbable because of the lack of testing and time from the point of re-design. We had surpassed the stretch goal and the foundation was set, demonstrating the Australian students once again can design and launch their own satellite successfully.

POST LAUNCH OPERATION

In the days following launch everyone was ecstatic about the launch and successful communication and seeking to learn more about how the satellite was operating. Going into the launch, we knew that the satellite would be power negative and so likely to move into a boot loop challenge where as the satellite started up, it would try communicate which would draw too much power and turn off the satellite again. We had a few days of strong communication and then silence.

Thanks to the SatNogs community hope is still there to fix this challenge. As the satellite is trapped in this bootloop process, we are now exploring the ability to turn on the satellite during the right orbit where it is drawing enough power from being in the sun over Germany. The orbit has to align as well with our ground station so that as it passes over Melbourne we can do an over-the-air firmware update to remove the function of communicating as ACRUX-1 boots up so that it can increase the power storage before attempting to communicate.

Plate 12.10

Plate 12.10 *The final satellite*

The MSP's vision is to launch the next generation of technology pioneers, those individuals who will shape the world. For us, ACRUX-1 has been the first demonstration of exactly that. We will continue to run more satellite projects and dive into other emerging technology fields to provide an incredible training experience that can develop the next generation of pioneers. A by-product of this will be establishing an ecosystem of engineers, scientists, entrepreneurs and lawyers not only prepared for the Australian space ecosystem but leading it. Through these projects we intend to support and develop a budding local industry and help Australia reclaim its place in space.

CHAPTER 13
Advance Australia—Where?

The chapter title is a reference to Australia's national anthem, 'Advance Australia Fair'. It's Richard's view on the history and the future of Australia's role in space.

* * *

Following the success of Australis, the question arose, 'What about Australis 2?' Why didn't the group go on to build another satellite, to start a business making small spacecraft, for a variety of uses? It would have created careers for the Project Australis team members, as well as many others, as later happened with Surrey Satellite, which grew out of enthusiasts like us, at Guildford University in Surrey, England. That company, now Surrey Satellite Technology Ltd, has grown to employ 485 commercial staff with annual revenues of £100 million and total annual export sales in excess of £600 million in 18 different countries. The founder, Martin Sweeting, has been knighted for his services to space technology.

But it was not to be. As Owen said in Kerrie Dougherty and Matthew Jones' excellent book, Space Australia:[1]

> We just couldn't face another one and we had dispersed and finished our degrees and went out and started learning what was not possible – what the real world looked like! … I guess in a way I regret now that, if I had been older and wiser, I would have started a business out of it because we had it all there. If we could have pulled it together as a space business, as Canada did for instance, then I think Australia would already have been one of the space faring nations of the world today, rather than aspiring once again to become one.

Tellingly, we did not even get a letter of congratulation from the Australian government for the Australis launch.

In order to understand why, in 1965, the year that we finished building Australis and took it to California to be launched, nobody had previously built a satellite in Australia, it is helpful to look at the history of making and launching spacecraft in this country. It is a story of decades of failed hopes and dreams, interspersed with moments of success and optimism.

In the late 1940s, in the aftermath of Britain's near defeat in World War II, the Woomera Rocket Range was established in outback South Australia, at the request of the United Kingdom government. The history of the range, and of the Weapons Research Establishment (WRE) at Salisbury, north of

---

1   Kerrie Dougherty and Matthew Jones, Space Australia; the story of Australia's involvement in space. Sydney Powerhouse Publishing, 1993), 54. The book has been being substantially rewritten and updated as Space In Australia for the 2017 International Astronautical Congress, held in September 2017 in Adelaide, Australia.

Adelaide, is described in Space Australia.

In an early attempt at a European space programme, ELDO, the European Launcher Development Organisation, was set up to build a satellite launching rocket to give Europe its own access to space, independent of the United States and the Soviet Union. The west coast of Cornwall was hardly the ideal blast off point. Far off Woomera, in the middle of a desert, was the obvious site to test the rocket, which was appropriately named Europa.

Europa was a mongrel – it had a British-built first stage, Blue Streak, which was a derivative of the American Thor missile, a French second stage, Coralie, a West German third stage, Astris, and an Italian-built test satellite. It was probably doomed from the start – the French hated the British, a feeling that was mutual, the Germans hated the French, and everybody hated the Italians.

Three attempts to launch test satellites with Europa from Woomera in 1968, 1969 and 1970 all failed to reach orbit – the last on 12 June 1970, six months after Australis' successful launch. Britain, in one of its many project cancellations of that time, withdrew from ELDO in 1968. The last Europa was launched from French Guiana in South America. It maintained the programme's perfect failure record. To my knowledge, Australia's only real contribution to ELDO was to provide the dirt on which the launch facilities were built in the South Australian desert. The sad remains of ELDO can be seen in the rusting launch pad at Lake Hart.

Earlier, in 1967, Australia was given the opportunity to build our own satellite. In 1966 ten Redstone/ Sparta rockets were shipped to Woomera from the United States in connection with atmospheric re-entry tests. As described earlier, Redstone was developed by Dr. Werner von Braun's World War II V2 missile team as a short-range ballistic missile. It had the honour of launching Explorer 1, America's first satellite, and the first Mercury sub-orbital flights, piloted by Alan Shepard and Virgil Grissom. Solid fuel rockets were added to the Redstone to drive the test vehicle into the atmosphere for the Woomera tests.

At the end of 1966, after nine flights, there was one Redstone/Sparta left over. This was offered to Australia to launch our own satellite, to be built by scientists and engineers at WRE. They, and the University of Adelaide in South Australia, already had experience in building payloads for sounding (that is, straight up and down, not into orbit) rockets. While everyone else involved was enthusiastic about the idea, the Federal government was not. They appeared unable to realise the scientific, technological and indeed the prestige benefits of the launch, and they initially refused.

Here the conspiracy theories start. Owen has already related the mysterious telephone call he received from a lady at WRE asking about the details of Australis and when it would be launched. That was a few months after we had taken Australis to California and we were awaiting a launch date – we thought before the end of 1967. The phone call showed that WRE, and therefore the government, knew that the scruffy University students had built the first Australian satellite and that its launch was approaching. Could that have changed the government's mind about approving WRESAT? We may never know, but we think it's a good story.

WRESAT was successfully launched into orbit from Woomera on 29 November 1967. WRE very kindly asked Project Australis (yes, they certainly knew about us) to send a delegation to watch the launch and, as Owen described, he and Peter missed the launch but Richard was there at Woomera when Australian space history was made. As a rather nice gesture, the day after Australis' launch in January 1970, WRE sent us a plate from their Baker-Nunn satellite tracking camera showing the ITOS-1 weather satellite and Australis, or Australis OSCAR 5 as it was then known, as streaks of light against the background of stars.

The authors of Space Australia said of the launch of WRESAT:

> The satellite success was a notable achievement for Australian scientists and engineers who hoped that its success would be the key to convincing the federal government to invest in a modest ongoing Australian space programme … Unfortunately, although the satellite succeeded in its scientific mission, it failed to provide the hoped-for impetus for a commitment to space from the government or the development of a national space policy.

There was simply no government interest. It was rumoured at the time that there were plenty of Redstones on offer at the US Army's arsenal in Alabama which could be made available for Australia's use to form the basis of a space programme at little cost. The offer was refused.

WRESAT was not to be the last satellite launched from Woomera – not quite. On 28 October 1971 some years after WRESAT, a British Black Arrow two-stage rocket successfully put the British built Prospero satellite into orbit. Of course, it had no impact on Australia, except that our desert was a safer and more politically correct launch site than the west coast of Cornwall.

The saga of FedSat well illustrates the lack of a plan for a space industry. FedSat was yet another ad hoc satellite project, a once off launch with no follow on. The Co-operative Research Centre[2] (CRC) for Satellite Systems was established in January 1998 with an eight year grant. It conducted research in areas such as satellite systems for communication, space science and computer technology.[3] Among the collaborators was the Institute for Telecommunications Research (ITR) at the University of South Australia where Owen served as a member of the Advisory Board of Management before and during the time of FedSat's construction. For more on the remarkable story of ITR, see the side box.

Part of the proposal for the CRC was that a satellite, FedSat, would be built and launched during the year of celebrations marking the centenary of the Federation of Australia. Now for those overseas

---

2   The CRC programme is 'a competitive, merit based grant programme that supports industry-led and outcome focused collaborative research partnerships between industry, researchers and the community', see https://industry.gov.au/industry/IndustryInitiatives/ IndustryResearchCollaboration/CRC/Pages/default.aspx and https://www.business.gov.au/assistance/cooperative-research-centres-program.

3   Encyclopedia of Australian Science, 'CRC for Satellite Systems. (1998-) at <http://trove.nla.gov.au/ people/1471320?c=people>. Accessed June 2017.

The Institute for Telecommunications Research

The Institute for Telecommunications Research (ITR) has had a stellar history in digital communications research, a discipline combining complex mathematics, electronics and computing. ITR's expertise was then in high-speed communications, demodulators for satellite ground stations, techniques for correcting errors in digital messages and it held the record for the closest approach to a theoretical limit for a communication channel (the Nyquist limit).

With this background, ITR became a member of the CRC for Satellite Systems funded by the Australian Government to build and launch Australia's first multi-experiment satellite – to be known as FedSat – to celebrate our nation's Centenary of Federation year in 2001. ITR undertook the design and construction of an advanced communications experiment package for FedSat. In addition, ITR's other challenge was to build and operate the telemetry, tracking and control Ground Station that would manage the satellite in its orbits around the Earth.

FedSat was successfully launched from Tanegashima in Japan on 14 December 2002. It continued to communicate with its University of South Australia Ground Station and permit operational experiments until August 2007 when it was turned off because its batteries were failing.

readers without a knowledge of Australian political history, "Federation" was the moment when the six Australian colonies of Great Britain came together to form an independent Australia on 1 January 1901, after having obtained the permission of the British parliament at Westminster.

Australia hadn't built a satellite since WRESAT and Australis in 1967, over thirty years before FedSat was thought of, so they were a little handicapped in spacecraft design and construction. Of course, we certainly could not launch it from Woomera – the last of the Redstones had decades before ended up in American museums and Prospero was the one and only satellite launched by the British Black Arrow as that programme was cancelled.

The structure and services sub-systems of FedSat were contracted to a British company which went bankrupt part way through construction. To their great credit, a "Can't-berra" company, Auspace, literally picked up the pieces and finished the job. The Japanese space agency, NASDA, now JAXA, very kindly gave us a free launch as a secondary payload on their H2 rocket.

Nevertheless, the critical and innovative parts of the satellite, the electronics, were made in Australia, in particular the communications package was designed and built by ITR.

So, finally, on 14 December 2002, FedSat was successfully launched into orbit. The fact that it missed the centenary of Federation on 1 January 2001 by nearly two years was swept under the carpet.

Now those criticisms would have been forgotten if the Australian government had grasped the opportunity, which they had failed to do after WRESAT, and used FedSat as the basis for a modest scientific and applications space programme. Alas, again, it was not to be. Following the announcement that FedSat would be launched in five year's time, I (Richard) wrote in a paper presented to a space conference in May 1997:

> *The fact is that FedSat is a stunt. It does not take five years to design and build a fifty kilogram satellite (in fact it took seven years). There are no funded plans for a follow on to FedSat and, while I support the building and launching of an Australian satellite, the fact is that FedSat looks like being another orphan, like WRESAT, with no presently planned programme to follow it. The launching of FedSat in 2001 will simply emphasise the gap of thirty four years since our last "official" satellite, WRESAT.*

My words were prophetic – there was no FedSat 2. Indeed it was not until 2017, fifteen years after FedSat's launch, that Australian-built satellites would again circle the earth. Three cubesats, compact enough to hold in the palm of your hand, built by universities in Adelaide and Sydney and a European consortium, were launched from the International Space Station in May 2017.

In 1971, again in 1974 and more recently, the European Space Agency (ESA) has invited Australia to join as an associate member, as Canada has done. Inexplicably, on each occasion the federal government declined, saying that it could not see any benefit in Australia being involved in space research! The ESA charter provides that contracts for the development of space systems are allocated to member countries in proportion to their contributions to the agency. ESA says that its members derive benefits of up to three times the amount that they invest in space activities. Australia, by its refusal to join ESA, has lost the opportunity to develop, over the past decades, a space industry for the country. These were fundamentally flawed and naive decisions, made by governments on both sides of politics, which at least demonstrates a bipartisan ability to get it very wrong.

The list goes on – the Aussat programme was established in the early 1980s to provide a domestic communications satellite service for Australia from geostationary orbit.[4] Regrettably, less than one percent of the three Aussat satellites and less than five percent of the later Optus satellites were built in Australia. The same applies to the two recently commissioned National Broadband Network spacecraft, Sky Muster, that were built in the United States and launched on the European Ariane rockets from French Guiana. This was because Australian space-related industries had little experience in manufacturing satellite components as a direct result of the lack of substantial local space activity.

There is a postscript to the Aussat story. NASA offered to fly an Australian astronaut as a payload

4   A geostationary orbit is one where the satellite's speed exactly matches the rotation of the earth and so appears to be stationary in the sky above the equator. The altitude of a geostationary orbit is 36,000 kilometres.

specialist on the Space Shuttle which launched the Aussat A-1 and A-2 flights in the mid 1980s. Of course, the offer was declined. Thus was lost a unique opportunity for Australia to be given first-hand experience of human space flight, to forge closer links with NASA and to create widespread Australian public interest in space research and development.

Owen added a comment here:

> I attended one of those NASA presentations where not only the offer of a flight for an Australian astronaut was made, but also the offer for Australia to develop LED cabin lighting for a proposed manned capsule. Imagine the flow on to LED lighting, including aircraft cabin lighting today! But, as Richard says, it was not to be.

Various organisations were set up by successive Australian governments over the decades to dabble in space research – the Australian Space Office, the Space Board, the Space Council, the CRC for Satellite Systems. Some of those resulted in hardware being built such as radiometers for European satellites and the Endeavour space telescope, which flew on the space shuttle. The University of Adelaide flew cosmic ray detectors on NASA's Pioneer 6 and 7 deep space probes, but there was never any follow on, no long-range plan that would have allowed an Australian space industry to grow and thrive.[5]

In the mid 1990s I was the Deputy Chair of the National Space Society of Australia and a member of the Executive Council of the Australian Space Industry Chamber of Commerce (ASICC). ASICC developed a plan for an Australian National Space Agency (ANSA) and presented it to the Labor government and the Coalition opposition ahead of the 1996 federal election. While the plan was well received, and it may have helped in the government's decision to go ahead with FedSat, the space agency did not materialise.

But does Australia really need a space agency and a modest space programme? The facts and figures are overwhelming. We are spending in excess of one billion dollars each year on space services and related activities when we could make and supply much of that in Australia. Those purchases contribute a significant amount to our balance of payments deficit while adding little to our technological or economic base. With an appropriate space programme many of those expenditures can be kept inside Australia which would not only decrease imports but materially assist in the expansion of our electronics, software and other space related industries. The cost of 'pump priming' an Australian space industry would be a fraction of that one-billion-dollar expenditure. Add to that the opportunities that are being lost for space science, engineering and other graduates from our tertiary institutions who must go to live and work overseas if they are to pursue careers in space research and development.

Then, at last, at the International Astronautical Congress held in Adelaide in September 2017, the Federal Government announced the creation of the Australian Space Agency. This was met with rapturous applause and a standing ovation from the Australian audience, many of whom had waited

---

5   Owen's comment: there are close parallels in the defence industry, especially naval ship building.

a lifetime to hear that news. The devil, however, was in the detail. The agency has a staff of 20 and a modest budget – certainly not enough to let contracts for the development of home- grown satellites, that would nurture an Australian space industry. But it's early days – we can only hope that the government funds and supports the Agency properly and that it does not go the way of its unsuccessful predecessors. Space related systems affect us all in our daily lives. Your mobile phone, the satnav in your car, bank Automatic Teller Machines, telephone and television, weather forecasting, farming, resource management and a myriad of other services rely on satellites for their successful operation.

And then, on the afternoon of Saturday, 29 June 2019 a cubesat, made by another group of space enthusiast students, at the same University of Melbourne, was successfully launched into orbit from New Zealand by Rocket Lab. Named ACRUX-1, after the brightest star in the Southern sky, the technical, legal, insurance and regulatory challenges faced by their team, 50 years after Australis, are set out in their chapter in this book. By comparison, the Australis team did not have to worry about such obstacles – nobody told us that they existed and our friends at AMSAT solved whatever they were. I like to think that a common bond between the Australis and the ACRUX teams was the joy that we felt when we heard that first HI coming down from orbit on the night of January 23 1970 – just as our 21st century successors felt when the first signals arrived from their little cubesat on that Saturday afternoon. Truly, the torch has been passed to a new generation.

Unlike Australis, the ACRUX team are planning to build further satellites – perhaps they will achieve what we did not – to start a business to benefit them and the country – and perhaps the Australian Space Agency will help them on their way. Australia has squandered many opportunities over the fifty years since WRESAT and Australis to establish a viable, economically productive space industry to serve the nation. As a result we are almost alone among the developed countries of the world in not having our own space programme.

It will take years, and a commitment of sufficient funds for the decades of neglect to be made up. However, there is no doubt that such a commitment is worthwhile and that it will result in a vigorous, innovative, high technology industry for Australia, which can lead to significant exports of space hardware, software and services within a decade.

While most Australians believe in space research and the benefits that it can bring to the country, the supporters of an Australian space programme have always had a hard road to follow. That road has been rocky and there have been few rewards over the past fifty years. However, history shows that those who have a vision, who have a goal to work toward and who apply themselves diligently to the task despite the many disappointments along the way, usually achieve that goal in the end, as did those scruffy nerds from the University of Melbourne who built Australis all those decades ago.

* * *

Owen's comments:

Australia does not seem to possess the "can do" attitude of the Americans of the era of Australis. I believe the lack of government interest in space is part of a wider community belief and culture that Australian technology and designers are unable to compete with overseas experts. This is, of course, a self-fulfilling belief – Australians can't do it, so we'll give it to a foreign outfit, so Australians don't. There are numerous examples and no doubt there's another book in it, too. It is also incorrect. As many will tell you, Australian engineers and scientists are among the most innovative in the world – we have to be to compete in a world of highly specialised people – and we tend to be generalists and perhaps see a broader view of an engineering challenge. Australians punch above our weight in Nobel prizes and the like.

Let us hope that the formation of the Australian Space Agency will encourage the development of an Australian space industry and, with it, the skills and intellectual property necessary for a modern high tech country.

CHAPTER 14
Postscript

If we think about the outcomes, two become clear: first the effects on our lives which were significant, at the very least. Both Richard and Owen found their life partners as a direct result of their participation in Australis: see the "Pauline Remembers" side box. Peter went on to achieve spectacular success with his work on OSCARs 6 and 7. Steve, Paul, Rod, Dave and the others also gained immeasurably from their involvement.

Secondly, we record with pride the firsts that we achieved with AO5:

- First Australian-built satellite. (WRESAT was built after AO5 but flown before).

- First non-US amateur satellite.

- First University satellite.

- First non-government secondary payload launched by NASA.

- First amateur radio satellite launched by NASA.

- First satellite, along with Tiros M, to be launched by the then-new Delta N rocket.

- First 10-metre HF transmitter flown on an amateur satellite.

- First telemetry system flown on an amateur satellite.

- First command system flown on an amateur satellite.

- First "PMASS" magnetic stabilisation and orientation system to stabilise and orient a satellite. There have been about 40 subsequent systems flown.

- First use of "Standard Orbits" method for predicting passes and for tracking. Owen was told that the method was later adopted and improved by NASA for use by US Antarctic expeditions.

- First time that satellite predictions (in this case equator crossings) had been transmitted directly from a computer to users via radio without human intervention.

In addition, AO5 trained two groups, the Australians and the US AMSAT group whose achievements have been noted.

It seems that Australis was among the pioneers of "small sats" at a time when satellites wee growing in size, weight and capability. Many, perhaps most, were launched on Delta rockets, as Australis was, from the Space Launch Centre at Vandenberg Air Force Base.

At the 23 July 2011 reunion of the group, Peter's contribution to AMSAT with his command receivers was acknowledged: the AMSAT 7 command system continues to operate, along with the RTTY telemetry, after some 40 years since launch, and it is the longest surviving space command system ever.

It was to be over eleven years before the Surrey University group under Professor Martin Sweeting[1] flew the first of their small and relatively inexpensive satellites.

As it turned out, we were pioneers but not intentionally so. We were simply following our interests.

May the younger generations of engineers, scientists and entrepreneurs do the same, for they are the real leaders of our technologically based society.

------

1   University of Surrey, 'Professor Sir Martin Sweeting' at <www2.surrey.ac.uk/feps/people/marytin_sweeting>. Accessed June 2017.

PAULINE REMEMBERS

I first met Richard at the house in which he and some friends lived in Drummond Street, Carlton. As Owen has described, I was invited as a "visiting cook" to prepare and cook the evening meal for Richard and his friends. Bizarre though that may seem today, it was actually an honour for a young lady to be invited to Drummond Street, Carlton not only to cook the evening meal, but to bring the meal with her.

Owen's version of the "satellite in the oven" is correct – there really were metal boxes with wires coming out of them inside the "Kookaburra" gas oven in the kitchen of the terrace house in Carlton. And Richard really did delay me cooking the meal until he had satisfied himself that he had completed the tests on the satellite parts.

About two months later, after Richard, Owen and Paul had taken the satellite to California and after I had returned from a Melbourne Symphony Orchestra tour to the United States and Canada, Richard arrived at one of the parties that my housemates and I threw. It was there that I noticed Richard, properly, for the first time – this nerdy but terribly handsome young man with Buddy Holly glasses. I experienced several terrifying rides on the back of Richard's motor scooter before he finally bought a car, a grey Toyota Corolla, in about mid-1968. Our romance flourished and we were married in January 1969, almost exactly a year before Australis was launched.

I supported Richard during the long wait to find out whether Australis was going to be launched or returned to Australia. However, on 23 January 1970, I joined the others in the MUAS control room at Melbourne University to listen to the successful launch of Australis. At last, we could all get on with our lives and our first child, Karen, was born in the middle of 1971.

Glossary

ABLS	Australian Balloon Launch Service
AO5	Australis Oscar 5
APT	Automatic Picture Transmission
ASIO	Australian Secret Intelligence Agency
Azimuth	The angle, measured clockwise along the horizon, from north to a point on the horizon[1]. See also Elevation.
BOM	Bureau of Meteorology, Australia
Case	Case Western Reserve University, Cleveland, Ohio
CIA	US Central Intelligence Agency
CRC	Co-operative Research Centres Program
Elevation	The angle measured from the horizon to a point above (or below) it. See also Azimuth.
FCC	Federal Communications Commission
GeV	Giga-electron Volts; that is, billions of electron volts
GPS	Global Positioning System
HF	High Frequency radio waves 3 MHz to 30 MHz (wavelengths from 100m to 10m)
IBM	International Business Machines company
IC	Integrated Circuits; that is, microelectronics
IGY	International Geophysical Year, June 1957 to December 1958
ITOS	Improved TIROS (q.v.) Operational Satellite
ITR	Institute for Telecommunications Research at the University of South Australia
JIB	Joint Intelligence Bureau, Melbourne, Australia
JPL	Jet Propulsion Laboratory, California
LSD	Lysergic Acid Diethylamide, a hallucinogenic drug favoured by hippies in the 1960s

Glossary continued

MASS	Magnetic Attitude Stabilisation System used on Australis
MUAS	Melbourne University Astronautical Society
NACA	National Advisory Committee for Aeronautics the precursor to NASA
NASA	National Aeronautics and Space Administration founded in 1958 assuming the responsibilities of NACA
NOAA	National Oceanographic and Atmospheric Administration
NSBF	National Scientific Balloon Facility
OSCAR	Orbiting Satellite Carrying Amateur Radio
PhD	Doctor of Philosophy
PMHG	Australian Post Master General's Department
RAAF	Royal Australian Air Force
RTL	Resistor Transistor Logic
RTTY	Radio Teletype
SLAC	Stanford Linear Accelerator Center
STEM	Science, Technology, Engineering and Mathematics
TIROS	Television and Infrared Observation Satellite
VHF	Very High Frequency radio waves 30 MHz to 300 MHz (wavelength from 10m to 1m)
VW	Volkswagen
WRE	Weapons Research Establishment
WRESAT	WRE Satellite launched on 29 November 1967

1   https://en.wiki.org/wiki/Azimuth for a more complete definition

Users Guide

The Australis OSCAR A Users' Guide is reproduced in full and as it was published in 1969. I have re-typed the pages and used a fixed spacing typeface reminiscent of the original typed document. Although I have originals of the photographs and diagrams, scanned images of one of the few remaining Users' Guides are reproduced for authenticity, complete with "bleed through" from the reverse side of each diagram page.


```
        AUSTRALIS  OSCAR  A



          USERS'  GUIDE



      *  *  *  *  *  *  *
```

The Australis OSCAR A Users' Guide contains full instructions for all wishing to track the satellite. Since the success of the project depends on the support of a large number of tracking stations,we are anxious to enlist the co-operation of suitably equipped radio operators, short-wave listeners and VHF enthusiasts everywhere. Further copies of this booklet are available from Regional Directors, or from:

> Project Australis,
>
> Union House,
>
> University of Melbourne,
>
> Parkville, Victoria,
>
> Australia. 3052.

Any enquiries or requests for more detailed information will also be welcomed by Project Australis.

CONTENTS

Frontispiece

Above: The completed satellite. The photograph shows the thermal
control pattern on the metal surface.

Below: A top view of the internal compartment. The battery
occupies the centre, and the electronics modules are
mounted on bulkheads at each side. The ends of the VHF
transmitting antenna, and the receiving antenna are
visible.

1. BACKGROUND

The Melbourne University Astronautical Society was formed at a time when the image of space research was dominated by a spirit of adventure. Today, much of the popular interest has subsided, but the potential of the spacecraft is being rapidly revealed. The satellite is an indispensible tool in many fields of research; its use in communications, navigation and meteorology is commonplace. The matter of communications, which received major publicity in 1962 with the success of Telstar 1, had already attracted the attention of amateur radio operators in the U.S.A.

At present the HF bands are overcrowded, but the traffic increases daily. One obvious solution is to move to higher frequencies. The early problems of noise and instability no longer haunt the VHF bands, but propagation characteristics severely limit the capabilities of VHF. Global communications may be achieved by such methods as moonbounce, but perhaps another solution is the artificial satellite. This has been accomplished, but still the amateurs are tied to the HF bands for international communications.

During 1965, the Melbourne University Astronautical Society began to investigate the problems of satellite construction. With the co-operation of Oscar, Project Australis was formed. Australis, like Oscar, aims to build communications satellites for use by amateur operators in all parts of theworld. In contrast to its American counterpart,Australis has no local background of satellite technology. This situation contributed to the difficulty in initiating the project. Financial limitations have also restricted progress. The result is that the first satellite is a relatively simple test vehicle, carrying two telemetry transmitters, a command system and a magnetic attitude control system. All electrical power is supplied by batteries which are expected to have an operating lifetime of about two months.

The satellite does not carry a repeater or translator.

It will be known as Australis OSCAR A until it is placed in orbit around the earth. Once in orbit, it will be given the next number in the OSCAR series to replace the "A".

The package construction, the command system, the antenna array and the magnetic attitude stabilization system could all be classed as experimental. The rest of the satellite provides the platform on which the experiments may be conducted.

However,when the experimental data must be recorded at a distance, the techniques of information transmission are added variables in the system.

For amateur operators and short-wave listeners there are opportunities to practise the art of tracking satellite signals in both the ten metre and two metrebands. The behaviour of the ten metre signal will illustrate long range propagation characteristics in the band.

In addition, there is a secondary objective. The project requires an efficient ground communications system to disseminate orbital figures and to collect data recorded by operators in all parts of the world. So far, the information channels have been organised,but the reliability of such a system has yet to be proved.

The final point illustrates the dependence of the project on human,as well as technical factors. Mechanical strength may be measured; electronic reliability has been improved with technology; for your assistance and co-operation we can only ask.

2. A TECHNICAL DESCRIPTION

2.1 Introduction.

The electronics of the satellite may be represented by the block diagram of figure 2.1. The physical layout is shown in figure 2.2. The obvious essentials are the two transmitters (10 metres and 2 metres) carrying the eight channel telemetry. To conserve battery power, a command system allows ground stations to control the operating time of the H.F. transmitter. A timetable will be published before the launch. A brief technical description of the spacecraft follows.

2.2 Hi Keyer

The hi keyer generates the Morse code identification. Although it operates continuously, producing the synchronization pulses for the telemetry encoder, its signal is transmitted for only 6½ seconds of each telemetry cycle.

2.3 Telemetry

Temperature, spin rate and battery performance are relayed to earth by the eight channel telemetry. Two temperature readings - one at the inside surface of the aluminium case, and the other from the insulated electronics compartment - are effected by thermistors.

Three phototransistors sensitive to reflected radiation from the earth are mounted on orthogonal axes. The output from each will indicate its orientation, and the rate of variation of all three is a measure of spin velocity. The channel sequence is:

0	hi identification
1	current drain
2	X axis horizon sensor
3	battery voltage
4	Yaxis horizon sensor
5	internal temperature
6	Z axis horizon sensor
7	skin temperature

The X, Y, Z, axes are as defined in figure 2.2 and figure 2.9.

In every case, the parameter is specified simply by the audio frequency. Unlike OSCAR 1 and OSCAR 2, the hi channel carries no telemetry data.

FLIGHT CONFIGURATION

COMMAND RECEIVER

COMMAND DECODER

VHF TRANSMITTER.

TELEMETRY.
CODING UNIT.

BATTERY PACK.

BASEPLATE

STABALIZA-
TION MAGNET.

HI KEYER

EJECTION SPRING WELL

HF TRANSMITTER.

Z

Y

X.

NOT TO SCALE

AXIS DEFINITION

AUSTRALIS OSCAR 'A' SATELLITE

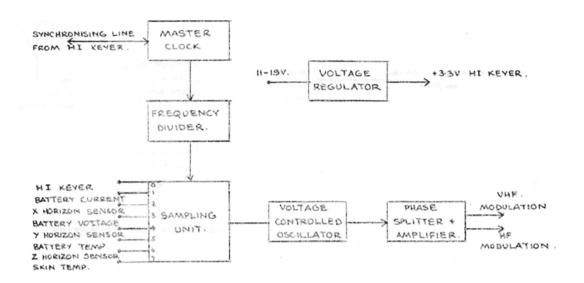

FIG 2·3　　TELEMETRY ENCODER　BLOCK DIAGRAM

$$HI = \bar{A} \text{ NOR } \bar{E} \text{ NOR } (\breve{D} \text{ NOR } [\bar{C} \text{ NOR } B] \text{ NOR } [C \text{ NOR } \bar{B}])$$

FIG 2·4　　HI　KEYER　BLOCK DIAGRAM

FIG 2·5   HF TRANSMITTER   BLOCK DIAGRAM

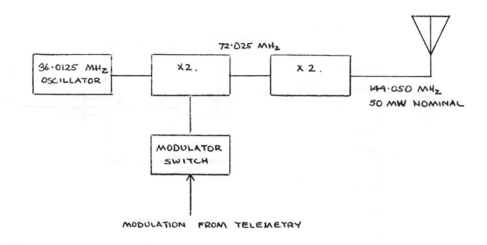

FIG 2·6   VHF TRANSMITTER   BLOCK DIAGRAM

A continuously operating switch ("encoder") samples each sensor for about 6½ seconds in the 52 second cycle. The voltage output is fed to an audio oscillator which modulates both transmitters. The audio frequency may vary from 400 Hz up to about 2,000 Hz. Graphs relating frequency to individual parameter values are given in chapter four.

2.4 V.H.F. Transmitter

A 50 mW crystal controlled transmitter operates continuously on 144.050 MHz. It is amplitude modulated by the telemetry.

2.5 H.F. Transmitter

The only ground commandable equipment is the 250 mW H.F. transmitter. It is crystal controlled on 29.450 MHz. The modulationis identical with the V.H.F. signal, except for a 180 degree phase difference. In each case the modulation index is 0.90.

2.6 Command System

Commands from earth are detected by a double change superhet receiver. The audio output is fed to the decoder which determines the validity of the command. When a correct signal is received, the decoder produces a control voltage to switch the H.F. transmitter.

2.7 Battery

Power is supplied by 28 alkaline manganese cells wired in two identical 20 volt series "strings". Each string supplies one transmitter, and the rest of the electronics run from both strings through an arrangement of protective diodes. If one string fails by short circuit or open circuit, then one transmitter is cut out, but the rest of the system operates. The diodes ensure that a short circuit in one string cannot impose an excessive load on the other.

2.8 Stabilization

To limit signal fading, and to maintain the antennas in a favourable orientation, some form of attitude control is necessary. Spin may be introduced at ejection,or by the prolonged action during thesatellite lifetime, of microscopic perturbing torques. The energy associated with spin is removed by magnetic hysteresis loss in an array of permalloy wires, and by eddy current loss in the aluminium alloy case. A bar magnet brings the X axis of the satellite into line with the earth's magnetic field.

FIG 2·7   COMMAND RECEIVER BLOCK DIAGRAM

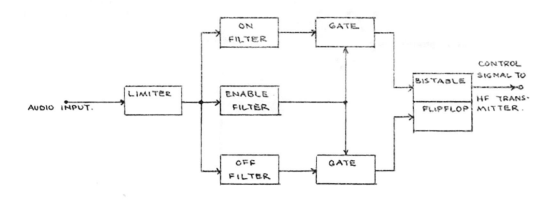

FIG 2·8   COMMAND DECODER BLOCK DIAGRAM

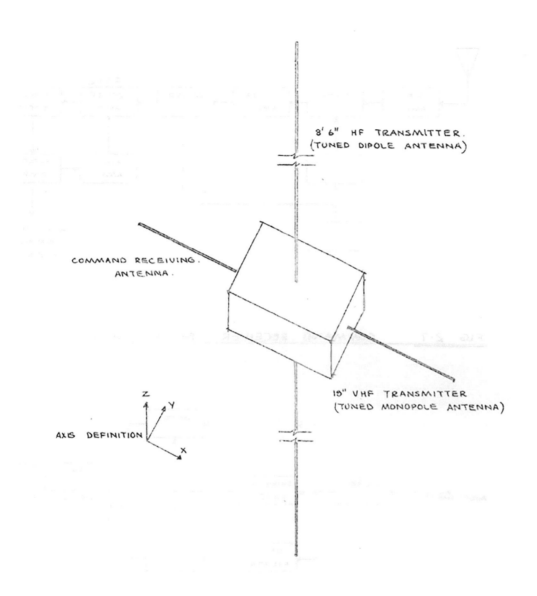

8' 6" HF TRANSMITTER.
(TUNED DIPOLE ANTENNA)

COMMAND RECEIVING.
ANTENNA.

15" VHF TRANSMITTER
(TUNED MONOPOLE ANTENNA)

AXIS DEFINITION

FIG 2·5 AUSTRALIS OSCAR 'A' FLIGHT CON-
-FIGURATION

2.9 Package

The electronics modules are mounted on an aluminium frame which is built around the battery compartment. A layer of thermal insulation separates all of this from the outercase. The aluminium alloy used for the case contains 1.0% magnesium,0.6% silicon, 0.2% copper and 0.2% chromium. A paint pattern on the outside surface is designed to maintain a fairly stable internal temperature by regulating heat radiation.

All antennas are made of flexible steel tape,and are mounted as shown in figure 2.9.

3. TRACKING INFORMATION FOR AUSTRALIS-OSCAR A

3.1 Regional Directors

For the purposes of disseminating tracking information, three regional directors have been appointed. Each director is responsible for distributing information within a specified area. When Australis OSCAR A has been launched, Project Oscar will obtain orbital data and distribute them to the regional directors who will send them to localco-ordinators. Local co-ordinators will complete the distribution to all tracking stations within their area.

Table 3.1 Areas and Regional Directors

North and South America	Project Oscar Inc., Foothill College, Los Altos Hills, California, U.S.A. 94022.
Asia and Australasia	Project Australis, Union House, University of Melbourne, Parkvilie, Victoria, Australia. 3052.
Europe and Africa	W. Browning,G2A0X, 47 Brampton Grove, Hendon, London, N.W.4, U.K.

3.1.1 Data Distribution Within Asia and Australasia

The local co-ordinators within the Asian and Australasian area act as links between the regional director and amateurs who are tracking Australis OSCAR A. The co-ordinator will have the following responsibilities:

(a) He will have equipment to provide two-way h.f. communication with the regional director for the reception of tracking information and the transmission of urgent data about the satellite condition.

(b) He will distribute orbital predictions to amateurs within his area,

(c) He will provide telemetry forms to tracking stations and return completed forms to:

> Project Australis (Telemetry),
> UnionHouse,
> University of Melbourne,
> Parkville, Victoria,
> Australia. 3052.

(d) The Users' Guide may be released to the press.

Table 3.2 lists the local co-ordinators for the Asian and Australasia

Table 3.2 Local Co-ordinators for Asia and Australasia

New South Wales
> A. Swinton, VK2AAK,
> P.O. Box 1,
> Kulnura,
> N.S.W. 2251.

Victoria
> W.M. Rice, VK3ABP,
> 34 Maidstone Street,
> Altona,
> Victoria. 3018.

Queensland
> L. Blagborough, VK4ZGL,
> 54 Bishop Street,
> St.Lucia,
> Queensland. 4067.

South Australia	B. Tideman, VK5TN, 33 Ningana Avenue, Kingspark, South Australia. 5034.
Western Australia	D. Graham, VK6HK, 42 Purdon Street, Wembley, Western Australia. 6019
Tasmania	P. Frith, VK7PF, 181 Punchbowl Road, Launceston, Tasmania. 7250.
Japan	Kenso Sano, JA1EC, 11-16 Misaki -2, Kofu, Japan.
Malaysia	C.W.C. Richards, 9M2CR, Telecommunications Training Centre, Jalan Gurney, Kuala Lumpur, Malaysia Cont'd.
New Zealand	B. Rowlings, ZL1WB, Mason Street, Onerahi, Whangerei, Northland, New Zealand.

3.3 Orbital Data and Predictions

3.3.1 Introduction

In order to obtain good v.h.f. telemetry records from Australis OSCAR A, it will be necessary to use moderately directive receiving antennas which must be pointed towards the satellite throughout the pass. This section describes the tracking data to be distributed by Project Australis and explain show to use it.

3.3.2 Using the Orbital Predictions

Throughout this section it is assumed that the satellite is in a circular orbit at a height of 500 statute miles, and with an inclination of 70 degrees to the equator.

Once the height and inclination of the orbit are known, the position of the satellite during a particular pass can be specified by the time and longitude of the previous north bound equator crossing of the satellite. The times and longitude of these northbound equator crossings will be predicted by Project OSCAR and distributed to local co-ordinators. A typical set of northbound equator crossings is given in table 3.3.

Each local co-ordinator will be provided with a set of standard antenna pointing angles, giving at two minute intervals, th esatellite azimuth and elevation angles and the number of minutes since the previous northbound equator crossing. These pointing angles will be supplied for a number of standard longitudes of the northbound equator crossing. A set of pointing angles for a standard pass is shown in table 3.4.

To obtain antenna pointing angles for a particular pass, choose the standard set which has a northbound equator crossing as close as possible to the actual longitude of the northbound equator crossing for the pass. This actual longitude will be given in the orbital predictions, such as those in table 3.3. Add the number of minutes given in the left-hand column of the set of standard pointing angles (table 3.4) to the time of the northbound equator crossing for the actual pass (given in the predictions, such as in table 3.3),obtaining the time for which the satellite is at the given azimuth and elevation angles.

ASCENDING NODES FOR AUSTRALIS OSCAR A

DATE	ORBIT	TIME	WEST LONGITUDE	
31JAN66	0000	0526	356	
31JAN66	0001	0707	20	
31JAN66	0002	0848	44	*************************
31JAN66	0003	1029	70	
31JAN66	0004	1210	96	

TABLE 3.3

STANDARD ORBIT CORDINATES

FOR STATION VK3ATM   MELBOURNE, AUSTRALIA,
    215DEG WEST,   37DEG SOUTH.

ASCENDING NODE 45DEG WEST

ADD MINUTES	AZIMUTH	ELEVATION
84	171	3
86	165	9
88	159	15
90	144	19
92	131	15
94	123	10
96	119	5

TABLE 3.4

THIS IS A SAMPLE COMPUTER OUTPUT

For example, if orbit number 0002 of table 3.3 is to be tracked at Melbourne, first obtain the longitude of the north bound equator crossing from table3,3 (44 W). Then choose the closest standard orbit, for which the longitude of the northbound equator crossing is 45W (shown in table 3.4). To give the actual time, add the equator crossing time to each time in the left hand column of table 3.4. Thus at 0848 GMT + 84 minutes = 1012 GMT the satellite azimuth will be 171 deg. and elevation will be 3 deg. The azimuth and elevation angles are similarly calculated every two minutes, giving the pointing angles shown in table 3.5.

Table 3.5 Calculated Pointing Angles for Orbit Number 0002

Time GMT	Azimuth deg.	Elevation deg.
0848+84 = 1012	171	3
+86 = 1014	165	9
+88 = 1016	159	15
+90 = 1018	144	19
+92 = 1020	131	15
+94 = 1022	123	10
+96 = 1024	119	5

As a rule, tracking stations will be able to observe two northbound passes about 100 minutes apart,followed about 12hours later by two south-bound passes about 100 minutes apart. This pattern will be repeated each day.

3.3.3 Schedules

As a rough guide, the equator crossing predictions are accurate for as long after issue as the satellite has been in orbit when the predictions are issued. For example, predictions issued three weeks after launch will be accurate for about another three weeks.

Each local co-ordinator will receive tables of Standard Pointing Angles and Northbound Equator Crossings as described below.

(a) Several months before launch,a set of Standard Pointing Angles for the expected orbit, and a set of typical Northbound Equator Crossings (for demonstration purposes only) will be issued.

(b) As soon as possible after the launch, a list of Northbound

Equator Crossings will be issued. This list will probably be accurate for only a few days. If the actual orbit is greatly different from that expected, a new table of Standard Pointing Angles will be issued.

(c) Throughout the satellite lifetime, lists of Northbound Equator crossings will be issued by both mail and amateur radio, sufficiently often to keep local co-ordinators well informed, probably at fortnightly intervals.

4. USING AUSTRALIS-OSCAR A

4.1 Introduction

Australia OSCAR A will transmit telemetry continuously at a frequency of 144.050 MHz, and at a frequency of 29.450 MHz when the transmitter has been commanded on. All tracking stations are requested to obtain telemetry data from either transmitter whenever possible, since telemetry reception and reduction is one of the major purposes of this project. The following sections give an outline of the minimume quipment needed to receive telemetry from Australis OSCAR A.

4.2 Receiving Antennas

4.2.1 V.H.F. Antenna

It is desirable to use a circularly polarised receiving antenna to reduce fading caused by changes in satellite attitude. This antenna should have a gain of at least 10 db. One suitable antenna is a crossed Yagi (two Yagi antennas pointing in the same direction, one with vertical and the other with horizontal polarization), one being connected through an extra quarter wavelength of cable, giving a 90 degree phase shift between the two driven elements. Another suitable antenna is a helix, such as the one described in "QST" for November, 1965.

To receive good signals while the satellite is at high elevations the antenna should be steerable in elevation as well as in azimuth. If measurements of the satellite spin rate are to be made, a horizontally or vertically polarized antenna should be used.

4.2.2 H.F. Antenna

If a linearly polarized antenna is used to receive the h.f. signal, fading will occur because of both satellite spin, and ionospheric Faraday rotation. Thus it may be difficult to determine the satellite spin using the h.f. signal, unless the operator is capable of separating the two variations, For reception of the h.f. telemetry, a pair of crossed horizontal dipoles, mounted one quarter wavelength above ground, will give a reasonably good omni-directional,circularly polarized pattern.

4.3 Converters

To obtain a good signal to noise ratio, the h.f. converter should have a noise figure of about 4 to 8 db. Most h.f. receivers should be adequate to receive the h.f. telemetry although some older receivers may need a pre-amplifier.

4.4 Receivers

Both transmitters are amplitude modulated, with maximum modulation frequencies of 2,000 Hz, so that receivers should have i.f. bandwidths of about 4,000 Hz. Except for initial acquisition of the signal, a b.f.o. should not be used, as the telemetry information will be lost.

4.5 Telemetry

Most of the information required about the satellite is derived from the audio telemetry, which has eight sequential channels. Each channel is transmitted for about 6½ seconds and the whole cycle lasts for 52 seconds. The hi channel consists of a 1.6 sec. tone, followed by a 1.6 sec. hi all repeated once again. The hi is transmitted not as MCW but as AFSK. Thus the tones do not key on and off, but switch between two tones of different frequency. The actual frequencies contain no telemetry information.

The hi channel is followed by seven tones, each 6½ sec, long and each sending information about one of the channels. By measuring the audio frequency and using the calibration graph for the channel, the quantity concerned can be determined. During telemetry decoding, the time should be watched carefully, as the frequencies of two adjacent channels may be similar and the transition from one to the next may not be audible.

The sequence of the telemetry channels is given in table 4.1.

Table 4.1 Telemetry Channels

Channel Number	Function
0	hi in Morse code
1	total battery current drain
2	X axis horizon sensor
3	battery voltage
4	Y axis horizon sensor
5	internal temperature
6	Z axis attitude sensor
7	skin temperature

Calibration charts for decoding the telemetry are given for battery current (figure 4.1), battery voltage (figure 4.2), and for internal and skin temperatures (figure 4.3).

To enable the telemetry reports to be evaluated by computer, all tracking stations are requested to enter their observations on a special telemetry coding form, as described in section 4.8.

4.6 Telemetry Decoding

One convenient method for decoding the telemetry is to use Lissajous figures. The received audio signal is applied to the vertical input of an oscilloscope and a sine wave from a calibrated audio oscillator is applied to the horizontal input. The frequency of the audio oscillator is adjusted until a stationary ellipse is seen, indicating that both frequencies are the same.

If the oscilloscope timebase has been calibrated, a set number of cycles can be displayed and the period of each cycle determined, and hence the frequency. If the timebase is free-running, as little syncas possible should be used to avoid changing the timebase calibration.

If an oscilloscope is not available, the frequencies of the received telemetry and of the audio oscillator can be matched by ear. Even with poor signal to noise ratios this method gives results accurate to within about 10 Hz at 2,000 Hz.

Another method, which in many cases can give better accuracy than any previously described, is to match the tone with a piano note. However, confusion of octaves must be carefully avoided.

Lastly, if the signal to noise ratio is good, the best method is to use a direct-reading frequency meter or digital counter.

If a tape recorder is used to record data, its speed should be accurate to within five percent, at worst, or else results will be seriously in error. Otherwise, operators are advised to practise measuring the frequency of an audio tone in less than seven seconds. It should be pointed out that inaccurate results are worse than none at all - an accuracy of at least ten percent is needed.

4.7 Readability and Signal Strength

The readability and strength of the received signal will be used in deciding the weight given to the decoded telemetry.

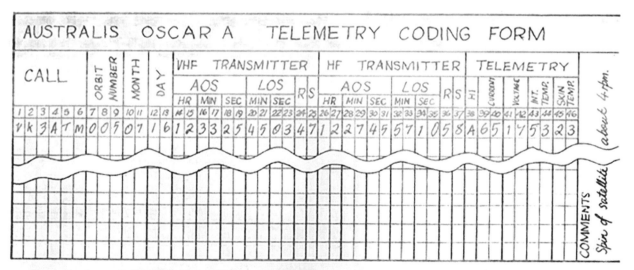

FIG 4·4   A TYPICAL PASS CODED - ALL SYSTEMS OPERATING NORMALLY

4.8 Telemetry Coding Form

Having decoded the telemetry for a pass, please select those results which you think are the most reliable. This will often mean rejecting wildly inconsistent results which may arise when the telemetry is decoded directly, rather than from a recording. Where a large,number of consistent results are obtained,all should be entered on the telemetry coding form, since this is an ideal indication of the reliability of the information.

Please write clearly, with only one character in each column.

All dates and times must be in GMT.

The following information is required :

(a) call sign of tracking station (if no call sign, write ZZ1, followed by the operator's initials)

(b) orbit number

(c) month and day

(d) time of acquisition of signal (AOS) and loss of signal (LOS), and readability and strength for each transmitter.

(e) hi keyer operation; the letter A for normal and F for failure, which should be described on a separate sheet.

(f) battery current drain in milliamp from figure 4.1

(g) battery voltage in volt from figure 4.2

(h) internal temperature in degree C from figure 4.3

(i) skin temperature in degree C from figure 4.3

All data entered on these sheets will be stored in a computer at Melbourne University. The form is in fact a replica of a computer card.

Reports on horizon sensor data should be treated differently. Since we are concerned only with either "light" or"dark",the actual frequency of the sound is of no interest. Each change in frequency corresponds to a transition of the field of view of a sensor between different states of illumination. The length of the higher frequency (bright) periods, depends on the spin rate,and on the nature of the traverse across the bright source. For example a short period could correspond to a single sweep a crossa short chord,or to a much faster sweep across a near diameter of the earth's disc. The sun and moon will also appear as bright sources against the dark background of space. However, they subtend such small solid angles at the satellite that the sensors will rarely sweep across them. Both would produce short high-pitched signals in the appropriate telemetry channels (Nos 2,4, or 6).

Now because the package may be rotating about three axes simultaneously, the spin rate on any single channel may not sound regular, except over a very long time. It is impossible to determine the spin rate directly. In fact it is a job for a computer, but this would require the recording of several telemetry cycles at various times. As far as individual operators are concerned, we would only expect a comment on the length of the sweeps across the earth. In this case, "fast" might be about one second; four seconds would be "slow", An average statement for each of the three axes is necessary.

Since computers do not take kindly to scientific information expressed in these terms,no columns have been provided on the telemetry form. A few words could be fitted in at the bottom of the sheet.

When the form is complete, please return to :

> Project Australis(Telemetry),
> UnionHouse,
> University of Melbourne,
> Parkville,Victoria,
> Australia. 3052

4.9 Station Details

Stations tracking Australis OSCAR A are requested to supply the following information about their station;

(a) name and postal address of operator

(b) call sign or station identification

(c) station latitude and longitude

(d) brief description of v.h.f. equipment such as antenna, pre-amplifier, converter and receiver

(e) brief description of h.f. equipment

(f) brief description of method used to decode the telemetry

Please send these details to the above address, and send emended information whenever a major change is made in your equipment, the date of the change.

Fig 4.1

AUSTRALIS OSCAR A TELEMETRY CHANNEL 1 CALIBRATION CURVE

Linear approximation to curve, $I(\text{in mA}) = \dfrac{f(\text{in Hz})}{9} - 63$

Linear region up to 1400 Hz.

Limit of the Linear Region →

BATTERY CURRENT (1 inch = 20 mA.)

TELEMETRY CHANNEL 1 FREQUENCY

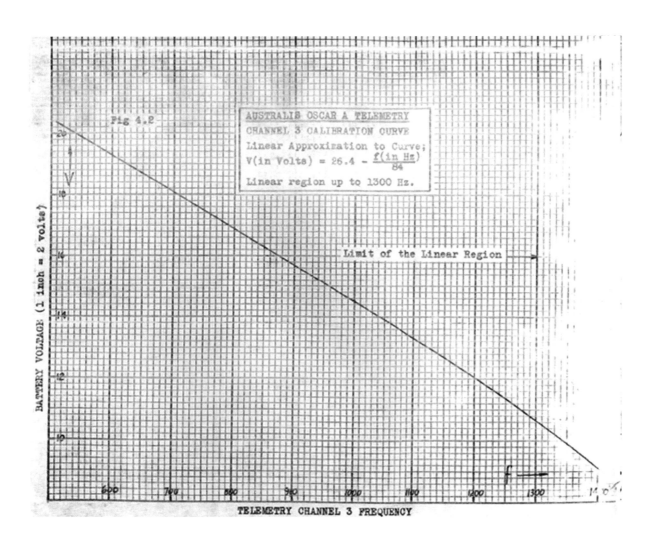

Fig 4.2

AUSTRALIS OSCAR A TELEMETRY
CHANNEL 3 CALIBRATION CURVE
Linear Approximation to Curve;
$$V(\text{in Volts}) = 26.4 - \frac{f(\text{in Hz})}{84}$$
Linear region up to 1300 Hz.

Limit of the Linear Region

BATTERY VOLTAGE (1 inch = 2 volts)

TELEMETRY CHANNEL 3 FREQUENCY

Fig 4.3

AUSTRALIS OSCAR A TELEMETRY CHANNELS 5 & 7 CALIBRATION CURVES

Linear approximation to curve; T(in °C) = 0.078.f.(in Hz) - 50
Linear Region up to 1200 Hz.

Limit of the Linear Region →

TEMPERATURE   (1 inch = 20°C)

TELEMETRY CHANNELS 5 & 7 FREQUENCY

Acknowledgments:

Project Australis gratefully acknowledges the kind assistance of the following organisations. Without their help the construction of the satellite would not have been possible.

Acme Engineering, Melbourne - RadioFrequency Connectors

Cannon Electric Ltd., Melbourne - ResistorsandConnectors

Ducon Condensers Pty.Ltd.,Sydney - All Capacitors used in the satellite.

Fairchild Australia Pty Ltd., Melbourne - All Semi-conductors used.

Melbourne University Union - A generous grant for ground equipment

Plessey Components Group, Sydney - One travel grant

The Potter Foundation, Melbourne - Travel Grants for two person to
                                    Project OSCAR

Pye Pty. Ltd., Melbourne - All Radio Frequency Crystals

Rola Co. (Aust) Pty. Ltd., Melbourne - MASS magnet

Sample Electronics, Melbourne - Circuit Boards

Turner Industries Ltd., Melbourne - Satellite antennae.

Union Carbide Australia Ltd., Melbourne - Flight and Backup Battery
                                    packs

Wireless Institute of Australia - A generous grant for running expenses

Thanks are also extended to the Meteorology Department of Melbourne

University and the Bureau of Meteorology, Melbourne, who have been most helpful during the construction of the satellite.

Australis-Oscar 5 Spacecraft Performance

Note From The Transcriber

This report was transcribed by me from a poor copy of Jan King's QST magazine article of December 1970. I have tried to transcribe the report as closely to the original as I can, including American spelling and two-column layout. The layout differs, of course, for a number of reasons. Tables II, III and IV have not been reproduced. Rendering by other software will no doubt change the layout.

Importantly, it includes references to other articles and papers.

Australis-Oscar 5 Spacecraft Performance

BY JAN S. KING,[1] W3GEY

---

1   c/o Amsat, P.O. Box 27, Washington, D.C., 20044.

In the rather brief lifetime of the Australis-Oscar 5 experiment a number of useful experimental and operational results have been achieved. The satellite was launched on January 23, 1970. As of this writing, 211 formal reports have been received from 27 countries around the world on both telemetry and propagation results. Many other stations were known to have received the satellite, but did not submit quantitative data.

Based on reports received, here is a summary of the performance of each system on the AO-5 spacecraft.

Thermal Behavior of AO-5

The temperature of AO-5 at ejection from the second stage of the Delta vehicle was 20 degrees C despite its proximity to the second stage engine and a very cold nitrogen gas jet during launch. The temperature, however, began to rise during orbits 1 through 10 and then stabilized internally at 43 degrees C ±3 degrees C where it remained for the duration of the

satellite's useful life. This temperature is fairly high, although it is within the design temperature range of 19 degrees to 45 degrees C. The effects of the higher temperature were, unfortunately, all adverse. Battery lifetime was somewhat shortened during the initial phase of discharge, but worse than this, the 144.05 MHz beacon power dropped off faster with decreasing supply voltage due to the decreased efficiency of the power output transistor.

External temperature measurements were higher in the sunlight and cooler during eclipse periods as observed by many responding stations. As the spacecraft entered the dark portion of the orbit the skin temperature dropped from its 55 degrees C average to 42 degrees C ±3 degrees C. The internal temperature, however, remained fairly constant, dropping only two or three degrees during the entire eclipse period. Acknowledgement is due Bill Armstrong, W0PGP, John Fox, W0LER, Nastar, K2SS and others for their data in this area. The spin rate about the X-axis in later orbits became quite slow so that the skin sensor located on the +Y surface showed changes in temperature as parts of the satellite rotated in and out of its own shadow. This data was most useful in determining the roll rate about the stabilized axis of the spacecraft. John Goode, W5CAY, reported this data for many orbits between 100 and 250. Skin temperature data indicated a spin period of 7 to 8 minutes about the X-axis after the initial 100 orbits. An example of this data is shown in Fig. 1 for orbits 168, 205 and 206 along with horizon sensor data.

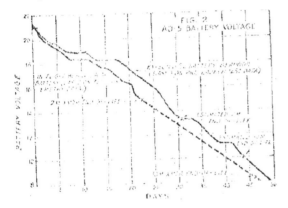

The AO-5 Power System

The spacecraft battery voltage decreased with time faster than predicted by pre-launch testing of individual cells (see Fig. 2). It is now known that the accelerated battery discharge was caused by two factors. First, the higher satellite temperature accelerated the normal chemical reaction in the alkali manganese batteries. Secondly, an additional 18 mA of current was attributed to a failure of the ten-meter modulator that occurred on orbit 3. It was verified that the 18 mA was independent of the ten-meter transmitter itself by commanding the transmitter OFF and observing that the extra current was still present. The ten-meter modulation failure has also been attributed to the higher spacecraft temperature.

The Magnetic Attitude Stabilization System and the Horizon Sensors

One of the best operating systems on board the satellite was not electronic in nature. The Magnetic Attitude Stabilization System (MASS) functioned more efficiently than some of us had anticipated. Early reports indicated that antenna nulls were occurring on the 144.06 MHz signal once every 15 seconds, making telemetry decoding very difficult. By orbit 100, signal fades had reduced to one or two per station pass (approximately 20 minutes in duration). To the station using the spacecraft this is a significant improvement over past satellites in the Oscar series and should prove to be a valuable tool in future amateur spacecraft to achieve the continuous reception of a down-link signal.

The three orthogonal earth or horizon sensors used in the spacecraft were 2N2452 photo transistors operated in diode mode, having a spectral response between 5400 and 10,500 Å. Each sensor's field of view had been stopped to 5 degrees by a small collimation tube. A photometric calibration of those sensors was unfortunately not undertaken due to the shortage of time in the test schedule. While the original design of this part of the telemetry system was to give an ON-OFF indication when looking toward or away from the bright earth, the devices were found to be more sensitive and capable of detecting the decreasing brightness of the earth's atmosphere as the sensors viewed the earth-to-space transition. When viewing the bright earth the telemetry output indication was approximately 1450 Hz and during the transition the telemetry frequency gradually decreased to a dark condition of 600 Hz. Amateurs using a fast discriminator to decode the modulation observed during periods of good signal strength small variations in the frequencies of the telemetry tones as the sensors swept across the earth's disc. These were attributed to cloud formations. Two examples of this data are shown in Fig. 3.

With a discriminator of this type, the Goddard Amateur Radio Club, WA3NAN, decoded telemetry information for all passes received. Figure 4 shows horizon sensor information for various passes. Each frame shows the maximum rate of change of brightness observed on any of the sensors during a given pass. During orbit 4 the maximum observed rate of frequency change was found to be 700 Hz per second, while pass 192 exhibits a maximum rate of change of only 10 Hz per second. This is indicative of the reduced spin rate of the satellite.

During daytime ascending nodes, after the spacecraft had stabilized, a regular sensor pattern was observed. W5CAY demonstrated this data most effectively (see again Fig. 1). The X-axis shows no true periodic nature, but rather a gradual transition followed by small variations about an average "light" condition. The Y and Z sensors show a periodic behavior characteristic of the satellite's roll rate about the stabilized X-axis. The skin temperature shows a cyclic variation as the +Y face rotated in and out of the spacecraft's own shadow. Of particular significance is to observe that the Z sensor always lags behind the Y sensor (approximately two minutes) in

detecting the earth. With the X-axis pointing north as the satellite crossed the equator , the spacecraft spin was thus clockwise as observed from the north pole of the earth.

The maxima in the external temperature curve were (within experimental error) out of phase with the +Y sensor. Since the T_{EXT} thermistor was located on the +Y face, then the temperature was a minimum during times when the +Y face was viewing the earth. This is, in fact, the time when the +Y face would have been in shadow.

As the spacecraft traveled north from the equator the +X axis should have begun to dip toward the earth as the strong dipole moment of the satellite (11,800 ???) followed the local geomagnetic field vector which caused it to rotate twice per orbit. (see Fig. 5). W5CAY's data showed that the +X sensor did begin to gradually come on shortly after the signal acquisition time over a period of several minutes. This is precisely what one would have predicted as the +X sensor looked deeper into the earth's atmosphere which reflected more and more scattered light into the sensor.

Fig. 1A Fig. 1B Fig. 1C

The average roll period observed in this data is 7.5 min. this is thought to be the degree of stabilization that persisted until the termination of the satellite's active life. The effectiveness of this system is best evaluated in terms of the very large reduction in the signal fading rate due to antenna nulls. This, in turn, implies an overall reduction in the loss of spacecraft data. For a satellite in the amateur radio service it is apparent that this method of stabilization is most effective and very easily implemented.

TABLE I

Region	Stations Reporting Useful Data	Stations Reporting Telemetry > 50% Passes	Stations Reporting Telemetry < 50% Passes
I	66	52%	48%
II	114	32%	68%
III	31	45%	55%

The A0-5 Command System

A telecommand link on two meters was utilized to turn ON and OFF the ten-meter beacon transmitter in an effort to conserve the spacecraft's power supply. An a-m tone modulation technique was employed. The ten-meter beacon which consumed 0.6 W of power, was to be commanded ON during the weekends when a maximum number of users was anticipated.

Prior to launch considerable difficulty was encountered with the spacecraft command receiver due to in-band interference from the 144.05 MHz beacon transmitter. It was only possible to eliminate the interference by adding a steep skirted bandpass filter centered at the command frequency. This filter gave 50 dB of rejection at the beacon frequency, but unfortunately had a relatively high insertion loss when placed in front of the receiver. The result was that the command receiver required a signal of -76 dBm (35.4 µV) under ambient (room) conditions to decode a command. This, to be sure, was considered marginal performance. The problem was further complicated

by the detuning of the second i-f stage that occurred during tests under vacuum conditions. The problem could not be traced to a single component in a timely fashion so it was decided to peak the receiver for maximum sensitivity under vacuum conditions. When the receiver was again tested under vacuum conditions the sensitivity was observed to be 10 dB better. Thus, it was expected that the in-flight sensitivity would improve some 10 dB over its ambient condition, giving a final sensitivity figure required to operate the decoder of -86 dBm. The spacecraft was launched with the receiver in this condition.

Figure 6 shows a plot of the spacecraft total current during the entire lifetime of the two-meter beacon, when telemetry data could be obtained. From this data it is clear when commanding occurred and the status of the ten-meter beacon during the lifetime of the satellite. Table III lists the command transmitter schedule indicating the successfully transmitted commands and the effective radiated power used to execute the command. Although early command attempts were unsuccessful, after orbit 72 it became increasingly less difficult to achieve a successful command and it became possible to maintain the weekend-only operation schedule for the ten-meter beacon as originally planned. It is felt that the increased overall sensitivity of the command system was due to a combination of factors:

a) Spacecraft command antenna orientation favorability (particularly over Australia due to the effectiveness of the magnetic attitude stabilization system.)

b) Reduction of the interfering signal level (144.05 MHz) as the battery voltage (and hence the power of the beacon) decreased.

c) Stabilization of the command receiver temperature and pressure which improved the sensitivity of the receiver.

The effectiveness of the command system, particularly despite the receiver problems, is of particular significance to future amateur space experiments. It not only demonstrated, for the first time in an amateur satellite, the effectiveness of ground command as a means of switching various experiments ON and OFF, but of greater significance, it represents an effective means of controlling amateur spacecraft emissions as to prevent interference to other services who may share amateur bands. This should help assure the continuing usage of amateur space experiments without the need for power flux limitations imposed on the spacecraft down-link.

Pass No. 4   Pass No. 54

Fig. 3 – Two examples of variations in the +Y sensor output due to variations in the earth's brightness. Note the sudden increase and decrease in intensity during the frame from pass 54. This is thought to be due to the sensor sweeping across a bright cloud region. Time divisions are 1 sec.

Spacecraft Lifetime

As previously indicated, the failure of the ten-meter modulator is considered responsible for the increased battery current drain of 18 mA. This additional current drain shortened the lifetime of the satellite. The two-meter beacon would be received through approximately orbit 280 on the 23rd day after launch. The ten-meter beacon was turned ON by command on orbit 261 and was left on continuously until it reached end of life around orbit 560 on the 46th day after launch. The difference in beacon lifetimes is due to the variation in cutoff voltage for the transmitters. The two-meter transmitter power output went to zero very rapidly at a supply voltage of 15 V while a significant output could be obtained from the ten-meter transmitter even at voltages as low as ten volts. While the spacecraft lifetime on two meters was shorter than the design lifetime of thirty days, a significant quantity of telemetry data was obtained never the less.

The Nature and Reliability of Amateur Reports

An additional feature of the AO-5 experiment was the opportunity to evaluate the performance of amateurs in reporting scientific-type data. After allowing several months to be certain that all reports had been received, an effort was made to determine what type of information amateurs were most interested in reporting and approximately how much variation in measurement occurred from station to station. It was decided to report on the results by ITU regions since different satellite passes were common to these regions (i.e., Region I (Europe and Africa)) could generally not hear the same passes as Region II (North & South Americas) and so forth). Table 1 lists the number of useful reports received from each region and those which did and did not contain telemetry information. We may infer that stations not reporting telemetry results were primarily interested in other aspects of the experiment or in phenomena such as Doppler measurements. (Only telemetry results are covered in this report since they were the primary indicator of the spacecraft performance. Another report by Raphael Soifer, K2QBW, gives a detailed presentation of the ionospheric propagation results of AO-5. Table 1 indicates that, on a percentage basis, Region

I and Region III participated more actively in the telemetry decoding activities. This is somewhat surprising, since it was anticipated that US amateurs would be suitably equipped to make telemetry measurements.

Fig. 4 The maximum rate of change of the horizon sensors during limb transition for various passes of AO-5. The data shows a despin factor of 70 in only 15 days. This is a particularly graphic demonstration of the effectiveness of the stabilization system. Time divisions are 1 sec.

It was of interest to determine the variation in measured values from as many stations as possible during a single pass. Variation in spacecraft parameters for a short period when the satellite passed over a region, was thought to be quite small (except for skin temperature variation) during daylight passes. The variation in data from reporting stations, then, can be primarily considered as individual station measurement error. In each region a particular pass was chosen for which a maximum number of reports was received.

Table II shows data for each station reporting and the range in data as well as the maximum percent of error from the median value. The error observed for the spacecraft battery shows the lowest error due to the relatively "flat" nature of the voltage-to-frequency conversion curve and the fact that most of those reporting rounded off their reported measurement (as called for by the telemetry reporting form). Certain stations (those underlined) were used as control stations for each region since they were known to have better than average decoding equipment.

All regions show comparable data error. The magnitude of the error (less than 10% max.) was approximately the error estimated prior to the launch. This data does not utilize more powerful statistical methods that could be used to more accurately evaluate the data (i.e., a uniform probability density was assumed for all data). The maximum error of 10% does indicate that amateurs throughout the world are capable of making significant data measurements with considerable accuracy.

Summary

With the exception of a failure in the modulator of the ten-meter beacon transmitter, all Australis-Oscar 5 mission objectives were met.

a) The spacecraft was effectively stabilized to two revolutions per orbit (geomagnetic alignment) within the lifetime of the satellite.

b) Reliable amateur spacecraft telecommand was demonstrated.

c) The effectiveness of the seven channel telemetry system was verified. Amateur data generally showed less than ±10% variation from median values.

d) Significant results were obtained on propagation effects over the satellite-to-earth link in the ten-meter band.

e) Partial success was obtained in achieving the design lifetime of several weeks for both spacecraft transmitters using only chemical batteries.

Fig. 6 – Motion of a magnetically oriented satellite in a polar orbit.

While the response to A0-5 was gratifying (many stations reported it to be the most interesting amateur space activity to date) it does not compare with the level of excitement that was generated by the repeater satellites such as Oscar III. Amsat is presently planning a next generation of Oscars. These satellites will carry two repeaters and an RTTY telemetry system capable of measuring as many as 60 parameters. The design lifetime of these satellites will be one year using a solar cell power source.

Whether you are interested in RTTY, fm, a-m, ssb, DX, traffic handling, or even contesting there are activities and special experiments being planned for you with Oscar 6. If you are interested in finding out how you can contribute to this new and exciting chapter in amateur radio, write to Amsat, P.O. Box 27, Washington, D.C., 20044.

Bibliography

Data taken from a series of reports on Australis-Oscar 5 submitted to Amsat by John Goode, W5CAY.

1. Data taken from Australis-Oscar 5 (A Summary Report) submitted to Amsat by John Fox, W0LER.
2. Data taken from Fairchild Semiconductor Specification information on 2N986/2N2452 NPN Planar Phototransistor, 5/62.
3. Information from preliminary data reduced at the Goddard Space Flight Center, NASA, by the Goddard Amateur Radio Club, 4/70.
4. Fischell, Robert E., "Magnetic and Gravity Attitude Stabilization of Earth Satellites," report CM 996, John Hopkins Univ. Applied Physics Labs, May 1961, p.38.
5. Op. Cit., John Fox, W0LER.
6. Soifer, Rafael, "Ionospheric Propagation from Australis-Oscar 5" (A Survey Report to the Radio Amateur Satellite Corporation). QSL October, 1970, pg. 54.
7. Op. Cit., Soifer.

Rocket, TIROS M and Orbit

Launch Vehicle

The launch rocket was a two-stage, long tank, thrust augmented Thor/Delta rocket[1]. In addition, there were six solid propellant, strap-on Thiokol Castor II motors used for the first time in this configuration with a Thor/Delta. Interestingly, three of the Castors were ignited thirty seconds after lift-off. The lift-off thrust was about 270 tonnes (nearly 600,000 pounds) and was capable of orbiting a satellite of twice the mass of the satellite it launched, TIROS M. Updated and improved versions are today manufactured and flown by the United Launch Alliance.

TIROS M Satellite

The TIROS M[2], also known as the Improved TIROS Operational Satellite (ITOS), was a cylinder 1.2 metres high, a span of 4.3 metres over its solar panels, and it weighed 330kg at launch.

The following instruments were carried: an Advanced Vidicon Camera System, an Automatic Picture Transmission (APT) system, a Flat Plate Radiometer, a Solar Proton Monitor, a Vertical Temperature Profile Radiometer and a Very High Resolution Radiometer. It provided data for nearly 18 months, ending its life on 18 June 1971.

The Orbit

We considered that the orbits of TIROS and AO5 were identical if for no other reason than that for several weeks we had no other means of knowing the orbit of AO5. In any event, it was a perfectly satisfactory assumption for the purpose of providing antenna pointing information for radio amateurs.

Apogee (highest point in the orbit):	1,479 km
Perigee (lowest point in the orbit):	1,434 km
Inclination:	101.8°
Period:	115 minutes exactly

---

1   See https://en.wikipedia.org/wiki/Thor-Delta; https://en.wikipedia.org/Thor_(rocket_family); and www.historicspacecraft.com/Rockets_Delta.

2   See http://www.astronautix.com/i/itos.html; and https://www.wmo-sat.info/oscar/satellites/view/203.

A Conspiracy Theory

Surely no book like this is incomplete without a conspiracy theory? Did the Commonwealth decide to go ahead with the WRESAT satellite in order to beat the students' satellite into orbit? Was there Australian pressure on the US Air Force to falter in its cooperation with Project OSCAR? The answer to both questions is almost certainly not, but …

Before examining any conspiracy theory, let us establish that Australis was indeed the first satellite built in Australia. To quote Des Barnsley, the project manager for WRESAT[1], 'Towards the end of 1966 it was clear that at least one of the 10 SPARTA vehicles would be surplus to project needs and the possibility of using it to launch an Australian designed satellite was unofficially raised. This idea quickly took off and agreement between the Department of Supply and the US DoD was speedily obtained.' Later in the referenced web page Barnsley states that 'Design work on WRESAT began at W. R. E. and the University of Adelaide early in 1967 ...'

Australis was delivered to the US in June 1967 while WRESAT was launched on 29 November 1967, soon after its tests had been completed. Australis was completed some five months before WRESAT and therefore earned the title of Australia's first satellite but not the first to orbit.

I have a number of newspaper clippings stating that we expected a launch of Australis within 12 months of being delivered to the US. It is ironic that WRESAT was launched before Australis.

I have a letter from the Minister for Air dated 4 March 1966 acknowledging my advice to him of our 'proposed satellite'. Thus the Australian government was aware of Australis nearly a year before WRESAT was approved and design work on it began. It is unlikely that the Minister for Air would not have mentioned Australis to the Minister for Supply who was responsible for WRE. However, I can find no evidence that WRESAT was approved in order to trump Australis nor have I been able to find any mention of WRESAT in the cabinet papers of the day.

Nonetheless, suggestive information, but not definitive proof, is the stock-in-trade of conspiracy theorists.

---

1 'WRESAT Background' at <https://www.honeysucklecreek.net/supply/WRESAT/background.html>. Accessed July 2017.

Image Credits

All images have been taken by the me and other members of the Australis OSCAR team with the exception of the following:

Plate 1.1: Credit: NASA.

Plate 2.1: Image Credit: Bauer Media Pty Limited.

Plate 2.2: Image Credit: National Archives of Australia. NAA: A1200, L85635

Plate 2.3: Permission of Arc @ UNSW Ltd.

Plate 7.3: Credit: Fairfax Syndication.

Plate 10.5: Credit: NASA.

Plate 10.9: Reprinted with permission from the University of Melbourne.

Plate 10.10: Credit: NASA.

CPSIA information can be obtained at www.ICGtesting.com
Printed in the USA
BVIW121350160520
579764BV00002B/14